The Essential Buyer's Guide

Ford Capri

All Capri models 1969 to 1986
(including USA Mercury Capri 1970 to 1977)

Your marque expert:
Mark Paxton

Other great books in this series, from Veloce –

Essential Buyer's Guide Series
Alfa GT (Booker)
Alfa Romeo Spider Giulia (Booker & Talbott)
BMW GS (Henshaw)
BSA Bantam (Henshaw)
BSA Twins (Henshaw)
Citroën 2CV (Paxton)
Citroën ID & DS (Heilig)
Fiat 500 & 600 (Bobbitt)
Ford Capri (Paxton)
Hinckley Triumph triples & fours 750, 900, 955, 1000, 1050, 1200 – 1991-2009 (Henshaw)
Honda SOHC Fours (Henshaw)
Jaguar E-type 3.8 & 4.2-litre (Crespin)
Jaguar E-type V12 5.3-litre (Crespin)
Jaguar XJ 1995-2003 (Crespin)
Jaguar/Daimler XJ6, XJ12 & Sovereign (Crespin)
Jaguar/Daimler XJ40 (Crespin)
Jaguar XJ-S (Crespin)
MGB & MGB GT (Williams)
Mercedes-Benz 280SL-560DSL Roadsters (Bass)
Mercedes-Benz 'Pagoda' 230SL, 250SL & 280SL Roadsters & Coupés (Bass)
Mini (Paxton)
Morris Minor & 1000 (Newell)
Norton Commando (Henshaw)
Porsche 928 (Hemmings)
Rolls-Royce Silver Shadow & Bentley T-Series (Bobbitt)
Subaru Impreza (Hobbs)
Triumph Bonneville (Henshaw)
Triumph Stag (Mort & Fox)
Triumph TR6 (Williams)
VW Beetle (Cservenka & Copping)
VW Bus (Cservenka & Copping)
VW Golf GTI (Cservenka & Copping)

From Veloce Publishing's new imprints:

BATTLE CRY!

Soviet General & field rank officer uniforms: 1955 to 1991 (Streather)

My dog is blind – but lives life to the full! (Horsky)
Smellorama! – nose games for your dog (Theby)
Waggy Tails & Wheelchairs (Epp)
Winston ... the dog who changed my life (Klute)

www.veloce.co.uk

First published in September 2009 by Veloce Publishing Limited, 33 Trinity Street, Dorchester DT1 1TT, England. Fax 01305 268864/e-mail info@veloce.co.uk/web www.veloce.co.uk or www.velocebooks.com.
ISBN: 978-1-84584-205-5/UPC: 6-36847-04205-9

Readers with ideas for automotive books, or books on other transport or related hobby subjects, are invited to write to the editorial director of Veloce Publishing at the above address.
British Library Cataloguing in Publication Data – A catalogue record for this book is available from the British Library. Typesetting, design and page make-up all by Veloce Publishing Ltd on Apple Mac. Printed in India by Replika Press.

Introduction & thanks
– the purpose of this book

In the competitive world of car manufacturing, image is a potent tool in convincing the public to part with its hard earned cash and buy into a dream artificially created by an advertising team, and woven around what is, after all, just another mass-produced consumer durable. It's a difficult trick to pull off, but, in 1969, the Ford Motor Company hit the nail firmly on the head with the launch of the Capri. The first sales brochure headline describing the Capri was: "The car you always promised yourself." This was no idle boast, and the public responded with delight to the new sporty, sleek coupé that provided the style and performance of much more exotic makes, but with the cosy reassurance of being based on well proven running gear that could be trusted not to claim the family's annual motoring budget at the first hint of a breakdown. The new car also boasted an unparalleled range of mechanical and trim options to complete an already attractive package. No other manufacturer at the time could offer anything close and, despite a few attempts in subsequent years, no one else could quite reproduce the same aura that surrounded the Capri.

It's surprising, then, that such a successful model could have languished in the classic car doldrums for so long. However, image is a double-edged sword, and the Capri's budget super car status took a hammering in the 1980s when it became the transport of choice for a more 'boy racer' clientele. Only now is it recovering from this association, and reclaiming its rightful spot as a milestone in the development of European motoring.

If having a Capri grace your driveway is a dream that you would like to make a reality, then this volume will take you through a buying process that will weed out the obviously unsuitable, then move on to a step-by-step look at the more promising cars. It will examine the most common frailties, both body and mechanical, concentrating mainly on UK built cars (though, happily, most of the information also applies to all models, including the US market Mercury Capri).

The unique points system included in the *Essential Buyer's Guides* will help ensure that you end up with the best example for your money, and your very own budget super car to cherish for years to come.

My thanks to the Capri Club International which allowed me access to its workshop for some under-body photographs, and to Rick whose much-loved Capri awaits restoration.

Mark Paxton

Contents

Essential Buyer's Guide™ currency
At the time of publication a BG unit of currency "●" equals approximately £1.00/US$1.50/Euro 1.20. Please adjust to suit current exchange rates.

1 Is it the right car for you?

– marriage guidance!

Tall & short drivers

The front seats will comfortably accommodate long-legged drivers, although the amount of room left for rear seat passengers may be severely compromised. Shorter drivers will have no difficulties as seat adjustment and travel are generous.

Weight of controls

These are heavier than more modern cars but not excessively so and appropriate to the Capri's image. Steering is light on the move, although more of a struggle at parking speeds. Most models benefit from servo assistance so the brake pedal requires only moderate pressure.

Will it fit in the garage?

The Capri is reasonably proportioned at 4.44 metres long, 1.7 metres wide and 1.32 metres high. The doors are large and need to be opened wide to allow easy entry or exit to the low slung seats.

Interior space

Despite being a supremely comfortable car in the front, some people may find it a little claustrophobic in the back, thanks to the low roofline and small back windows, which are fixed on some models, but even in pop out form, do little to help alleviate the feeling.

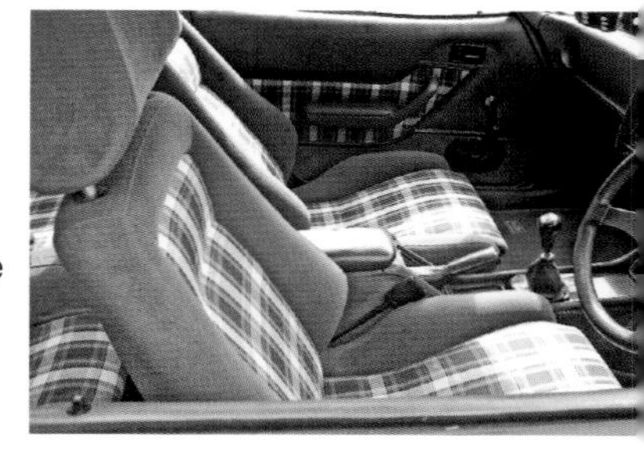

The Capri cabin is comfortable and welcoming.

Luggage capacity

Hatchback cars have folding rear seats, which extend a reasonable 9 cubic feet space into a pretty generous area of 22 cubic feet. Cars with a traditional boot are obviously more restricted, but adequate for two people's holiday luggage.

Running costs

Capri style comes at a pretty affordable price. Fuel consumption is average for the smaller engined cars, although a V6 may prove wallet-busting when it's time to visit the pumps. Servicing and repair costs are kept in check thanks to the family connection to lesser models in the Ford range.

1300 Kent	29mpg
2000 Pinto	26mpg
3000 V6	20mpg

Usability

1300s will struggle in modern conditions thanks to sluggish acceleration and an ever present feeling of the motor working hard at higher cruising speeds. 1600s, whether

Pinto or Kent, are perfectly acceptable as a daily driver as long as you do not expect to beat modern hatchbacks away from the lights. The 2-litre cars are better still, and although still not rocket ships, provide a sufficiently sprightly performance to be interesting. The V6s are pretty potent but their strength lies more in the easy-going way the power is delivered, rather than the extent.

Parts availability

Mechanically, there are few problems sourcing parts, although bits for the UK-produced V4 can take a little time to track down. Many spares are still freely available from local motor factors, and often at low prices. Body parts are a problem, especially for Mark 1 cars.

Parts cost

The Capri is probably one of the cheapest cars with sporting pretensions to keep on the road, although body restoration can be costly thanks to the scarcity and expense of most body panels. Trim bits are also thin on the ground and therefore command a premium. It's definitely better to avoid cars with tatty interiors as they will be disproportionately expensive to sort out.

Insurance group

Classic car insurance is probably the best option if you can live with the mileage restrictions. If your Capri is to be used daily check with the owners clubs for the best company.

Investment potential

The Capri has been undervalued for a long time. Prices are rising now though and, as the number of available cars drop, the better ones will appreciate accordingly. It's unlikely to be a car that will ever suddenly rocket in value, but definitely one that will provide a high "smiles per mile" ratio without worrying about depreciation.

Foibles

Very few, as the Capri was a carefully designed car based on well proven predecessors. The back end can be twitchy in the wet with that situation becoming more pronounced as engine sizes increase and as rear suspension bushes and springs wear with time.

Minus points

Rust is a constant enemy of Capri owners and has consigned thousands to a premature end on the scrap heap.

Plus points

Reliability, simplicity and good engineering; plus what other car can offer that level of style at such an affordable price?

Alternatives

None. Other manufacturers tried to launch sporty coupés based on family saloons but none really hit the spot like the Capri. Even Ford found it impossible to emulate the car's success with subsequent models.

2 Cost considerations
– affordable, or a money pit?

Small service ● x60
Large service ● x140
Reconditioned engine Pinto ● x1100
Reconditioned engine V6 ● x1750
Reconditioned cylinder head (unleaded) ● x265
Reconditioned gearbox 4-speed ● x385
Reconditioned gearbox 5-speed ● x450
Reconditioned prop shaft ● x195
Brake caliper (exchange) ● x65
Radiator ● x100
Water pump ● x45
Pads ● x18
Shoes ● x24
Brake drum ● x45
Brake disc ● x25
Rear wheel cylinder ● x18
Wing front Mk1 Pattern (steel) ● x350 (when available)
Wing front Mk1 (fibreglass) ● x110
Wing front Mk3 Pattern ● x90
Shock absorbers Full set (4) ● x230
Reconditioned steering rack (manual) ● x45
Reconditioned rack (power) ● x155
Track control arm ● x26
Complete professional car restoration ● x6000 plus
Full professional respray ● x2000

Parts that are hard to find
Trim bits for all models are a problem to track down, even second-hand, as are genuine Ford body panels.

Parts that are expensive
Most mechanical bits are relatively inexpensive, although the cost of repair panels needed to restore some cars can mount up. Re-manufactured wings and other panels for Mk1 cars.

A replacement engine, new or second-hand, shouldn't take much finding.

Suspension and brakes are cheap to buy, and easy to source.

3 Living with a Capri

– will you get along together?

Most people will find the Capri easy to get along with, unlike some classics, mainly due to Ford's sensible decision to base the car on mundane running gear from its family saloon range. Highly strung and temperamental were never going to be descriptions that this car would be burdened with, and although reliable, sturdy and dependable may not be adjectives to set an enthusiast's pulse racing, they're much sought-after attributes if you do not want to spend all your spare time and cash trying to keep some metal primadonna in fine fettle.

Ford knew exactly the type of owner that the Capri would appeal to, namely someone who appreciated a little style to enjoy once the weekly commute was done and dusted. Now that the car is firmly ensconced in the classic sector, it provides exactly the same rewards thirty years on, although it's unlikely to have to endure the daily grind anymore.

With the demands of regular work lifted from its shoulders, even the smallest engined variant can be a satisfying car to own, as long as you accept the inevitable conflict between show and go that accompanies that choice. Move up to a V6 and, although outclassed by many modern family cars on paper, in the real world the Capri's willing torquey-ness can still upset a few current rep-mobiles, and do it accompanied by a glorious, throaty growl from a pair of pipes unrestricted by the demands of pollution control. On an A road the Capri is pretty much in its element, although protracted motorway trips at speed may prove tiring in earlier cars as soundproofing was not the refined science we are used to today, and anything with a four speed box may have you reaching for a higher ratio that sadly isn't in there. Cabin comfort is otherwise assured across the range thanks to a relaxed driving position with wide comfortable seats, which are adequately bolstered to keep you in place when pushing on through the twisty bits. The ride is reasonably modern, too, at least on good roads, and the controls are light enough that you do not have to worry about developing unsightly muscles in odd places, unlike drivers of more traditional sporting cars, so it should never really be a chore dragging the old girl out for even the longest journeys.

Although sporty, a Capri with a hatch is very practical.

When the time eventually comes for repairs or restoration, most versions offer a warm welcome to the home mechanic, with a generous engine bay allowing good access to oily bits which seem staggeringly simple today. The body, although rot-prone as we shall see during the evaluation process, is as straightforward as the mechanicals. Some have described the construction as crude, but, once again, that means that there is little stopping an enthusiastic owner from getting stuck in. That situation extends to improvements and upgrades too, as, even today, there are plenty of tuning and bodywork options available to personalise or enhance your car.

Panels can be expensive, which is pretty much the only downside of Capri ownership.

If the prospect of getting your hands dirty doesn't appeal, then the Capri shouldn't frighten the local garage either, as it's easy enough for it to be handed over to the apprentice to sort out: should anything out of the ordinary crop up, there will usually be an old hand still around who cut his spannering teeth on Fords like these. The bill shouldn't be too hard to swallow at the end of it all, and it's unlikely that a Capri will ever spend much time off the road waiting for an elusive part. In fact, given the poor stock levels held for modern cars at main dealers, you may find the situation better than if you owned a new model.

The final plus point in the Capri's armoury of charm is its ability to turn heads. That shape draws attention, as a long bonnet and short tail still epitomises speed and style in the minds of both petrol heads and the general public alike.

As a package, there are few classic alternatives that can lay claim to such a list of attributes, or blend them in a manner that appeals to quite so many people. As long as you can avoid a rusty example, which is where this book comes in, living with a Capri should be nothing but a motoring pleasure.

The car has an active and enthusiastic following around the world.

4 Relative values

– which model for you?

This chapter will attempt to give an idea, expressed as a percentage, of the relative values of individual models in good, useable condition. Continental and US cars may not follow the same pattern as those in the UK, so consult current magazine price guides for an up-to-date valuation in those markets.

The Capri has a fairly compressed range of values at the moment, with the rarity of earlier models unable to command the higher prices often found with other marques. The RS models, which are so far ahead of the ones listed here in value, and so rarely available, have been excluded from this list.

Mk3 280 Brooklands 100%

A combination of end-of-the-line limited edition, high specification and usability, plus being the youngest and, theoretically, the least worn out and rusty, all combine to put the Brooklands at the top of the value pile. Probably the ultimate expression of the Capri. Beware of fakes, however, as it's not unknown for lesser cars to be disguised to increase their appeal and worth. Do your homework on fixtures, fittings and chassis numbers, and speak to a club expert before parting with your money.

The last of the line and the most expensive are the 280 'Brooklands.'

Mk3 2.8i 60%

Like the Brooklands, age and usability are in its favour, with the added bonus of being more widely available. You can save a lot of money and still enjoy the Capri experience as long as you can forego the cachet of a limited run.

Mk1 3-litre GT 55% (E plus 5%)

Purity of the original concept, combined with useable power, even on today's roads, is a compelling mix. The higher spec trim and extras of the E command a premium.

Mk1 3-litres are fine cars with great period charm.

Mk2 & Mk3 3-litre 55%
You are likely to pick up either version for roughly the same amount of money as there seems to be little differentiation between the models in the eyes of the buying public. The choice probably boils down to styling differences and the fact that the Mk3 cars are much easier to find. Once again, the appeal of the larger motor helps values, though that may change with the rising cost of fuel.

Mk2 versions are only slightly less valuable than their predecessor; both are rare.

Mk1 GT 1600/2-litre 50%
Pretty rare today, these limited edition models were fitted with high spec interiors and had the power bulge bonnets normally only found on 3-litre cars at that time. Although these differences may not seem sufficient grounds for enhanced value, the cars seem to be in demand and command a better price than non-GT models of the same age and condition.

Mk1 1600GT was a well respected car in its day.

Mk1 1600/2-litre 45% (plus 5% for 1300 due to rarity/novelty value)
The 1600 offers the perfect blend of reliability and power, which reflects Ford's engineering excellence of the time and would lay the foundations for the Capri's future success in the showroom worldwide.

The 2-litre V4 has the appeal of more potent power which balances the less desirable engine.

Mk2/3 1600/2000 40%
Mk3s of these engine sizes are probably the most common Capris still around, and definitely the easiest to live with thanks to good parts availability, reasonable running costs, and cheap insurance. Performance levels are

A Mk2 1300 like this is astonishingly rare, and will command a slightly higher price than a 1600 in the same condition.

still high enough to provide the motoring experience that the Capri was designed for. In reality, the price gap to the Mk1s of the same engine size is pretty small, so, once again, styling and youth may play a big part in the decision.

Special editions

There were several attempts by Ford to add spice to the range by making special editions of certain models. These cars will be worth little, if any extra over their standard counterparts unless there is a buyer with a specific interest.

The Laser started out as a special edition but was so popular it became a mainstream product.

Restoration projects

These cars are of low value due to the high cost of getting a Capri back into good shape, as a result, they're worth only around one tenth of the value of a similar car in good condition unless there are compelling reasons to pay more.

5 Before you view
– be well informed

Once you have found a car that may be of interest, it might be worthwhile to consider some, or all, of the following points, which will help determine whether a particular vehicle is worth pursuing.

Where is the car?
The days of your local paper having dozens of Capris in its classifieds are, sadly, long gone, but there should still be several on offer within a reasonable distance. If you are after something a little out of the ordinary, then you will most definitely have to budget time and money to go and view the car you want.

Cost of collection & delivery
If you are buying a roadworthy, legal Capri, then you can simply drive it home after ensuring that you have arranged appropriate insurance. If you have bought a restoration project, costs will rapidly mount if you have to rely on someone else to move it for you. Always try and get a quote for this before you view your prospective purchase, and incorporate the cost in your negotiations.

Dealer or private sale?
There are unlikely to be many Capris on mainstream car dealers' forecourts as they're pretty much the preserve of specialists now, although they do appear in classic car dealers' showrooms. Expect to pay more for cars sourced this way, as the price will have been set to include overheads, guarantees and advertising, but at least you will have some comeback if anything goes wrong. If buying privately, make sure you see the car at the address on the V5 (ownership papers), and avoid anyone who uses a mobile phone only and offers to bring the car to you, or wants to meet at a car park or petrol station.

Reason for sale?
If the car is being sold by a specialist, then the answer is quite simple; it's their living. Private vendors, though, may have a much wider range of reasons, most of which are perfectly valid, but always ask: the question still takes some people by surprise, and a hurried answer might raise some doubts.

When & where?
Always view a car in daylight and preferably not when it's raining, as a damp sheen can make even the worst paintwork look half decent, as does fading light. Once again it's worth making sure that the viewing address matches that on the ownership papers.

Condition
Before you view, ask the seller for an honest appraisal of vehicle condition, tell them you are coming from some distance away, which usually helps, as they're likely to at least own up to the blatantly obvious defects. If it doesn't sound very promising there are plenty of other Capris out there, so don't waste your time.

All original specification?
The Capri has historically attracted owners willing to spend time and money adding performance parts. How successful this has been may be open to debate in many cases and, today, cars in totally standard trim are worth most, unless the modifications have been done by a recognised specialist.

Matching numbers?
The chassis number can be found on a plate on the slam panel, or driver's side inner wing, depending on age. Post 1981 cars have the number stamped in the body next to the driver's seat, accessed by a small flap in the carpet. USA market cars have the number on an additional plate visible through the windscreen. The vendor should be able to confirm that numbers match the paperwork before you arrive but make sure to double-check them yourself once you get there. Altering identities is pretty simple with cars of this age, so if you are looking at a high value variant make sure that you do your homework on the correct fixtures and fittings.

Is the seller the legal owner?
In the UK the ownership papers record the registered keeper of the vehicle, which may not be the actual owner. Ask the vendor if they're the legal owner, if not, get some contact details of the person who is and check that they're aware the car is being sold.

Is the vehicle taxed & MoT'd?
All countries have some sort of roadworthiness check; in the UK it's referred to as the MoT. Inspect the certificate to see how long it's valid for, and make sure the registration and chassis numbers match the car you are looking at. Do not rely on the presence of a valid certificate as a guarantee of the car's condition; it only reflects the state of many components at the time of testing and in the opinion of the person doing the examination. There are a few comebacks if the test has been incorrectly carried out, or if it's suspected of being illegally obtained, but it is better to check the vehicle thoroughly yourself and avoid heartache later. In the UK, road tax for a 1300 Capri is in the lowest bracket, the rest pay the higher rate, although that may change in the future as older cars are demonised for their emissions. Cars built before 1973 are currently road tax exempt in the UK. Other countries have similar concessions for older vehicles.

Unleaded conversion?
All Capris, apart from the very end of production, cannot use unleaded fuel without an additive or having the cylinder head modified.

How can you pay?
There is no doubt that a wad of cash can be a useful lever to get the price down when you are negotiating. If you are unsure about carrying a large amount around with you, a personal cheque could be left, although collection would probably have to wait until it had cleared, or you could use a banker's draft drawn on your account in the vendor's name. Many sellers are now wary of these, however, as fraudulent versions are not uncommon. Dealers are usually the most flexible when it comes to payment methods, and will normally accept credit and debit cards.

Are you insured to drive if you buy?

There are several ways to cover yourself to drive your newly purchased Capri home, many using a policy initially arranged for other vehicles. This may be attractive in the short term as it saves money, but should there be an accident, the amount of cover provided would probably not equate to the value of the car. Classic policies are cheap, and not necessarily that restrictive if you shop around, so make sure you are adequately insured before hitting the highway.

Professional vehicle checks

Many motoring organisations around the world offer this type of service for their members, and should have no trouble examining a Capri. It's also possible to get a specialist to do it for you, as the cost may be less and they should be well used to all the car's idiosyncrasies. Whoever does it, make sure they have liability insurance so you can make a claim in case they miss anything which proves to be expensive later.

Data checks

Once again, national motoring organisations can help find out the history of your car, whether it has been stolen, written off, or is subject to an outstanding hire purchase/loan agreement. In the UK the best known organisation is HPI – 01722 422 422. Such organisations usually provide a compensation scheme in case any of the information that they supply is inaccurate. Other countries have similar organisations.

6 Inspection equipment

– these items will really help

This book

The major defects that you are likely to find in any prospective purchase are outlined in this book. It would be all too easy to forget important checks in the excitement of viewing a new car, but if you keep this volume to hand and methodically work your way through the sections, using the unique marking system, it will help you make up your mind, and could save a lot of your hard earned money from being wasted.

Magnet

Take a weak one; if it's too strong it will cling on through the layers of filler that you are seeking to discover. When you use it, don't scrape it along paintwork, apply and remove it with care. Remember that the car you are looking at is someone else's pride and joy until you stump up the cash to make it yours.

Overalls

Sadly, Capris rust, which means rolling about on the ground to find out the extent of the rot. A set of disposable overalls can be discarded after the inspection so you can take a test drive without sullying the car's interior and irritating the seller.

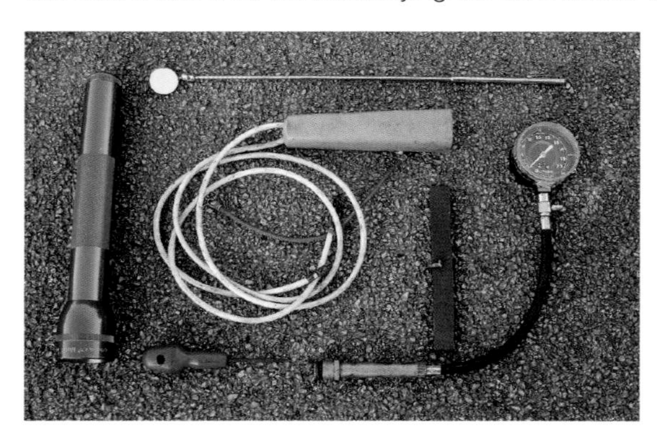

The bare minimum inspection kit.

Torch

There are a few hidden nooks and crannies in a Capri body, a torch will make both under-bonnet, and underbody inspection easier, even on the brightest of days.

Spectacles (if you need them for close work)

Most cars have had some filler at some time, and, unless it has been applied with extra care, there will be small telltale marks under the paint, so take your specs if you need them for close-up work, as a detailed look at some areas will be needed.

Jack & axle stands

The Capri is a low slung car so a trolley jack will help lift it sufficiently for a more comfortable look at the underbody and steering/suspension. Never get under a car that is supported by a jack only; ALWAYS use stands.

7 Fifteen minute evaluation

– walk away or stay?

Having found a possible purchase, asked all the necessary questions over the phone, and assembled your inspection equipment, you should now find yourself confronting a Capri in the metal. The most important thing is to make sure that your heart doesn't rule your head, so keep calm and consider just how long you have had to work to save the money (usually a sobering thought). It should be possible to eliminate obvious heaps very quickly, hence this 15 minute examination, by the end of which you will know if the car warrants further time and effort, in which case, some of these areas will be revisited and checked in greater detail. It's important that just as much care is taken when looking at a 1980s Capri, as they can rot just as badly as older versions.

Let the light reflect off the sides to show up any defects.

Exterior

First impressions count, but don't fall for shiny paintwork, look instead at how the car sits; there should be no signs of sagging, particularly at the rear where weak springs are fairly common, and in some cases these may even be broken. Continue the general assessment by looking down the length of the car, tilting your head until the light catches the panels which will throw poorly filled areas into sharp relief. The Capri is fairly slab-sided so any bodged repair work should readily show up. Pay particular attention to the wings (fenders), the bottom of the doors and the rear wheelarches. Check around the front and rear screens for flaking, bubbling, scabby paint, or any attempts to disguise rust. Lift the screen rubber and see if there are signs of the car having had a re-spray with the glass left in: check the scuttle carefully whilst you are at it.

At the front, have a look all over the wings as rust can erupt at the trailing edge, around the lights, behind the wheel, at the join to the front and slam panel, the lower valence and the inner wings. Replacement panels are expensive and hard to find for earlier cars, so expect to find filler in any, or all of the locations listed, but at the moment limit yourself to a general impression.

The scuttle rots, so expect to find filler or blistering paint.

The front wings corrode badly, especially at the front.

Sills are a weak point.

Check the rear wheelarch all the way around.

The door step suffers in the corners but can be rusty along its entire length.

The back panel and rear valence hole and split.

Be critical of the strut tops, even when repaired like this one.

Look closely at the door gaps for any signs of unevenness, then open and shut each one in turn; they should do so cleanly, any sign of them lifting up and on to their strikers as they close, or dropping as they open, is potentially a big problem, and, once again, will warrant further investigation. Look, too, at the top of the doorstep which often corrodes: is it still in the original paint or has it been undersealed? Have a look at the inner sills (rocker panels) if possible, although, in many cases, they may be covered in carpet.

If that is the case try squeezing the box sections through the covering, badly rusted ones can be felt and heard giving way under the pressure. Bend down and check the outer sills – you are looking for any sign that they have been patched, filled or blacked out with more underseal. If they're still in their factory paint, look for signs of rot bubbling through. Check that the jacking points are still there, as they often rot away completely.

At the back end check the boot (trunk) lid for signs of rust, particularly at the bottom; the same applies to hatchbacks, which can suffer under the spoiler, too. The rear valence may be rusty and split at the bottom under the bumper and where it meets the back wings, which should be easy to spot. Open the tailgate (if fitted) and lift the false floor to check for holes in the corners, left and right, or signs of accident damage. Check the outer lip of the rear wheelarch for filler or other indications of corrosion.

Lift the bonnet and have a look at the inner wings, in particular, the area around the strut tops. Look for rust staining, fresh underseal or paint. If there are any welded repairs, do they look neat? Has a proper repair panel been used, or home-made patches? Are there signs of rippling or any other damage?

Interior

Fortunately, the Capri is pretty basic inside, so its defects are easy to spot. The seats often sag, the covers are not particularly robust, so split easily, and even the frames come apart at the seams. The door trims are usually just card covered in vinyl, which can warp and split from a combination of age and damp, problems which the carpets suffer from, too, and wet carpets mean rusty floors. The instrument panel is fairly simple, even on the best equipped versions, but check for any additional holes left by a previous owner's attempts to 'upgrade.' Have a look at the headlining whilst you are in there as it's tricky and very expensive to replace.

Mechanicals

The engine bay is unlikely to be pristine unless you are looking at a recently restored car, but it shouldn't be an oil-congealed, dirty mess either – an indication of a general lack of care. Pull out the dipstick and check the oil level, it should be near the top mark and the lubricant should not be black and thin. Wipe a little on the end of your finger and roll it around for a second to see if it feels gritty, then smell it, very cheap stuff can give off a faint aroma of ammonia, and old oil can smell burnt; neither is good. Look for signs of a creamy brown

The state of the oil can tell you a lot about an engine's condition.

emulsion on the stick which could indicate head gasket problems. Any water staining on the block, under the join to the head, the core plugs, around the water pump and at hose connections are never welcome.

The coolant should be clean, without any scum or rusty sludge.

Take off the radiator cap (or expansion tank depending on age and model) and check the state of the coolant. It should be clean and brightly coloured, depending on make and chemicals included; you do not want to see scummy, orange rusty water, or a creamy emulsion like you checked for on the dip stick, which again would indicate a head problem. Have a quick look at the radiator for damaged fins or signs of leaks.

Ask the vendor to start the engine after you have positioned yourself where you can clearly see the exhaust tailpipe. The motor should fire up promptly from cold (and if the car was hot when you arrived, ask why), and it's common for combustion to be accompanied by a short puff of smoke as a small amount of oil that has crept past worn valve stem oil seals burns off. The important thing here is that the smoke should clear very quickly indeed, if the car continues to produce a blue haze after more than a few seconds, there is piston or ring wear, particularly if it increases with engine speed. Black smoke when revved in short bursts means that the mixture is too rich, and white or grey smoke indicates the presence of water, which may clear once the car is warm; if it doesn't once again, the head gasket is the primary suspect.

Finish off your fifteen minutes by listening to the engine; does it idle evenly and quietly? Are there any knocks and rattles? Are the tappets clattering away? If they are it could be a sign of poor maintenance or a dry top end, which indicates oiling problems, irregular changes of lubricant or, if it's a Pinto motor, a worn camshaft and followers. Does it look standard under there or have things been altered; if so, why?

Is it worth staying for a longer look?

The answer, of course, depends on the price being asked for the car and the level of expectation attached to that price tag. If you are looking to do a complete restoration you may not regard any of the above as a problem, although it's best to avoid cars with extensive corrosion unless confident of your abilities with a welder, and are fully aware of the cost of the repair panels that you will need to complete the work. If the car is being sold at around the market price for that model in good condition, then finding dropped doors and lots of filler may be much more of an issue. If things seem okay so far, or at least not so bad that you have been put off, then it's time to move on to a far more searching, and critical examination; so get out the overalls, you're going to get dirty!

8 Key points
– where to look for problems

Strut tops
This area is under tremendous load from suspension forces being fed into the shell. Any deterioration spells trouble so make sure that the whole section above and below the wing is free of rust and not distorted. Repair panels are available and cheap, but to do the job correctly takes time and some skill. Poor repairs will do nothing for the strength of the car. Get in really close and look for signs of any bodges designed to disguise the real state of the metal.

Sill structure
Like all monocoque cars, the Capri relies on its sill structure for strength. New outer panels are inexpensive but must be fitted correctly to regain full rigidity. Inner sills are often patch repaired, so make sure that any new additions are securely welded around the full circumference, and that the welding is to a good standard. Uneven, lumpy runs sitting proud of the surface are not good enough.

Rear spring hangers
A tricky area to repair successfully and, like the strut tops, must be done well as the shell takes the full force of the rear suspension at this point. Unless the car is spectacularly cheap or has some sort of sentimental value, it's best to move on and find another one if this section is badly corroded.

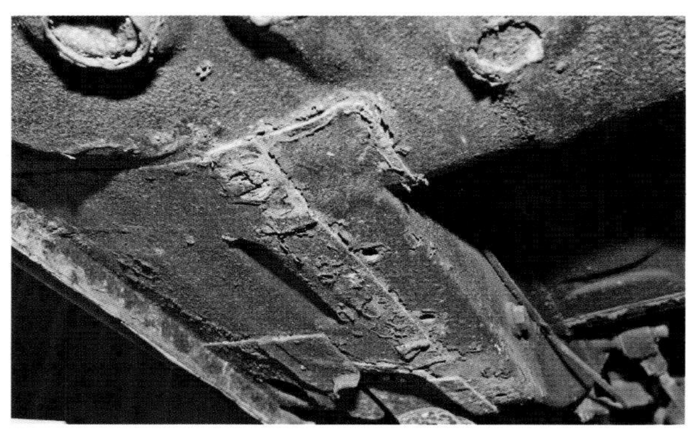

Dropped doors/A-post

The A-posts and the side of the bulkhead rot badly, allowing the large and heavy doors to drop down. At the bottom this usually combines with inner and outer sill rot to create a festering mess. It's all repairable but takes considerable time, which means money if you are paying someone else to sort it out. Be critical of the door gaps and how they open and close.

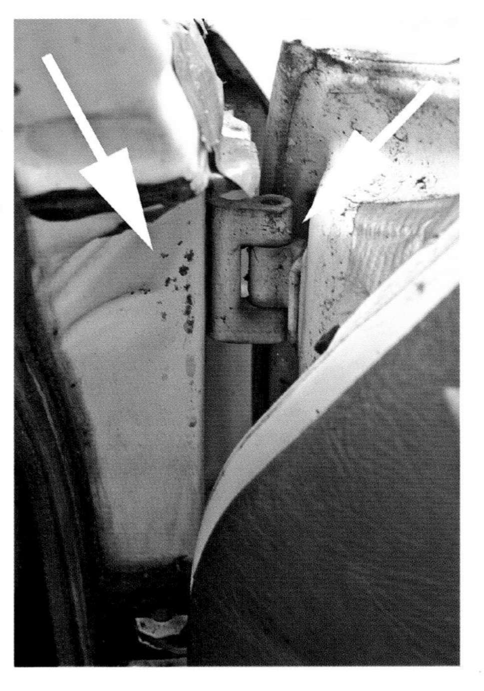

Damaged interior

This is one area which is often overlooked when checking out a classic, but it cannot be ignored on a Capri as replacement seat covers, headlining, door cards and even carpets can be almost impossible to find, and will be a big drain on your bank account if you are lucky enough to track some down. A tatty interior can be a deal breaker unless you can live with it and the price reflects that attitude, as future resale should always be kept in mind.

9 Serious evaluation

– 60 minutes for years of enjoyment

Score each section as follows: 4 = excellent; 3 = good; 2 = average; 1 = poor

Exterior

Spotting filler

Before we dive into the nitty-gritty of a full blown inspection, it may be worth thinking about how you are going to spot any filler that may have been used to disguise the condition of your prospective purchase; unless it's a low mileage, well cared for car, it's going to be there somewhere. The text and pictures will help identify where you are most likely to find it and, unless it has been applied with extreme care, there will always be some clues to its presence. Look through the paint for sub-surface scratches or swirls left by a dual action sander or coarse abrasive paper. Often tiny holes will have remained unnoticed until the paint went on and not considered worth the trouble or time of redoing the job. If you have any doubts look to the panel edges, as the repairer will have tried to lose the filler there, well away from the initial damage and far enough for care levels to drop; this is especially true under the lower curve of the body, at the bottom of wings or sills. Use your hands to gently move over suspicious sections, as the sense of touch is an incredibly useful tool for detecting imperfections. You cannot hope to root out all applied filler, particularly if it has been done professionally, but run-of-the-mill Capris have rarely had that level of attention. The final tool in the battle against the bodger is knowing how the car should look, for example where seams and spot welds are visible, which is where your pre-purchase homework of looking at as many cars as possible will prove time well spent.

Paintwork

In the 15 minute examination shiny paint was deliberately ignored to concentrate on more gritty issues, but now is the time to assess just how good the paintwork really is. If the car has been resprayed very recently it may well have been done to cover up lots of bodged repairs so that the owner can shift it at a healthy profit. If that is the case, the work may well have been done on a tight budget, so look for signs of a rushed job; for example, overspray on the window rubbers and seals, or an orange peel finish or dull, flat sections where there is an inadequate depth of paint. If it looks like it might have been messed about with, pull out that magnet you brought with you, it's going to see some action as you work your way around. If the finish looks original and carries normal battle scars and dents, it may well have faded, and if the surface has gone 'milky' it could be beyond saving, no matter how much time and T-Cut you lavish on it. With resprays now costing very large sums of money it could be a budget buster if you need your Capri to be shiny.

Look for signs of overspray on rubbers, around handles and glass.

Front wings

4 3 2 1

Check the area around the headlamps carefully, both on top of the wing and down the side, as trapped mud rots out this section with monotonous regularity. The rear trailing edge blisters and holes down its entire length, a couple of inches in from the door gap and the bottom section behind the wheel suffers too, often parting company from the sill, leaving the wing flapping. The join to the front valence splits due to corrosion and the arch lip needs examining, as parking scrapes rapidly rust. Expect more rot at the top rear corners next to the scuttle as well. Wings are expensive and awkward to fit as the door has to come off, and that requires dash removal to access the bolts. Expect to find filler, or other cheap repairs, as a result. Non-genuine wings can often be a poor fit so check panel alignment carefully.

Have a look at the join to the sill; this one has self tapping screws holding it together.

Front wings are a real weak spot.

Inner front wings

4 3 2 1

Revisit the strut tops making sure that everything is as sound (or otherwise) as your initial impressions suggested. Look for signs of distortion in the metal, around the mounting bolts and surrounding area, or attempts to disguise the beginnings of rust. If the tops have been plated they should be seam welded completely around the join, and show no signs of swelling or other disfiguration. Shine your torch in the wheelarch and look at the underside of the wing, especially the plates which form the cup and the

The cup and side plates corrode and split away from the inner wing.

Take your time checking the strut tops and look for signs of underlying rot.

Battery trays hole.

side supports that the strut sits in. These should be in one piece and show no sign of separating from the inner wing, or from each other. If it's a V6 Capri there should be an additional plate welded in here which often rots off, compromising the shell's strength. Be very suspicious of new underseal in this area or any sign of seam sealer.

Whilst under there check the join to the bulkhead panel behind the wheel as this suffers from having road debris thrown at it; later cars may have arch liners which will help protect it but make inspection difficult.

Finish off the inner wing inspection by checking the join to the outer wing inside the engine bay as this is a prime spot for rust, then make sure that all the flat sections are sound, particularly where they join the bulkhead and front panel. At the bottom a chassis box section is clearly visible; ensure that it, too, has not been attacked by corrosion and that there are no signs of rippling or other impact damage. Do not forget the battery tray, as it can rust badly.

Bonnet

The leading edge rusts out, as do the front corners, so check for signs of bubbling under the paint. The corrosion actually starts on the inside so expect worse when you look at the lip behind. Hinges seize and can cause the bonnet to flex which, in turn, can cause corrosion to start so have a good look in the area.

The box section on the inside of the bonnet rots out first.

Check around the bonnet hinges, too.

The slam panel to inner wing join splits, and this is often disguised with seam sealer.

The front valence loses its bottom edge.

Front panel/slam panel/front valence   

The front panel corrodes badly around the headlamp mounts, but check the section under the radiator grille as that goes, too. The slam panel corrodes where it meets the inner wings, and check the bottom of the front valence which becomes 'frilly.'

Open the doors and move them up and down, checking for movement in the A-post.

The side of the bulkhead can rust badly at the bottom.

A-posts   

This is a real problem area and will be the source of major expense if you miss the rot. Open the doors, lift them up and down and look and feel for movement. There will be some in the hinge pins but if excessive the whole post may move. Check the top of the panel for holes, especially if you found rust damage at the back of the wings and around the scuttle, as water will have had access to the rear of the A-post with predictable results. The first sign of this can be staining seeping through the seam. Get into the car and use your torch to look up into the dark recesses under the dash, and check the state of the inside join to the bulkhead.

Windscreen surround and scuttle

4 3 2 1

This whole area is susceptible to corrosion, so take your time to double-check that any rot has not been disguised with filler. The bottom corners and the scuttle suffer the most.

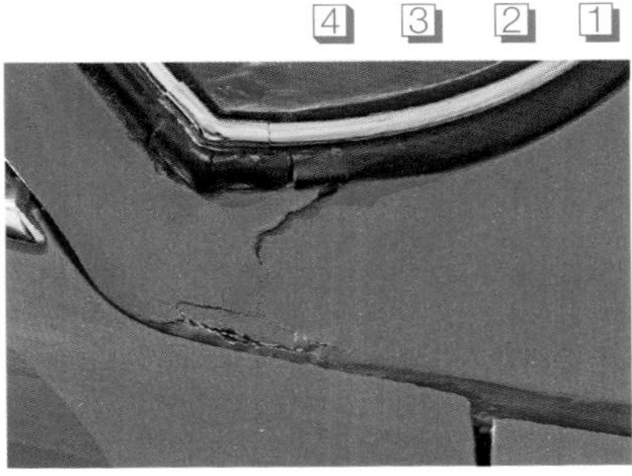

Scuttle rot is often filled.

Sills

The sill structure is a crucial part of the shell's strength so must be in good condition. If the outers have been replaced it must have been done properly. The front and back ends are the most likely starting point for trouble, so look carefully for bubbling paint; if you find any give the area a good hard push. Be suspicious of stone chips or underseal as filler may well lie underneath. Part repairs, by way of small plates welded in, may be sufficient for the car to have passed an MoT (annual safety inspection), but are only a short-term solution and proper replacement will not be far off. Skin sills were popular when the Capri was just another second-hand car and, once again, may be fine for a quick test result but actually do little for the strength of the structure, so try and determine if the front of the sill continues behind the front wing or if it stops short. The bottom lip needs careful assessment in particular as it's often crumbly, so look for irregular lines or lumpy, drooping welds. If you squeezed the inner sills during the 15 minute evaluation and had any doubt about their condition, try and lift the carpet for a better look, although it may be securely trapped by trim strips, or glued. The lower section of the inner sill can be checked from underneath later on.

Sills rust at the back and along the bottom edge.

Inners go, too. This one has been extensively patched.

Floors

4 3 2 1

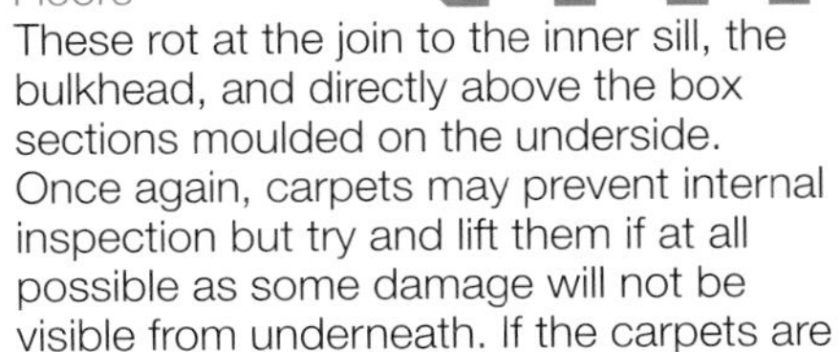

These rot at the join to the inner sill, the bulkhead, and directly above the box sections moulded on the underside. Once again, carpets may prevent internal inspection but try and lift them if at all possible as some damage will not be visible from underneath. If the carpets are

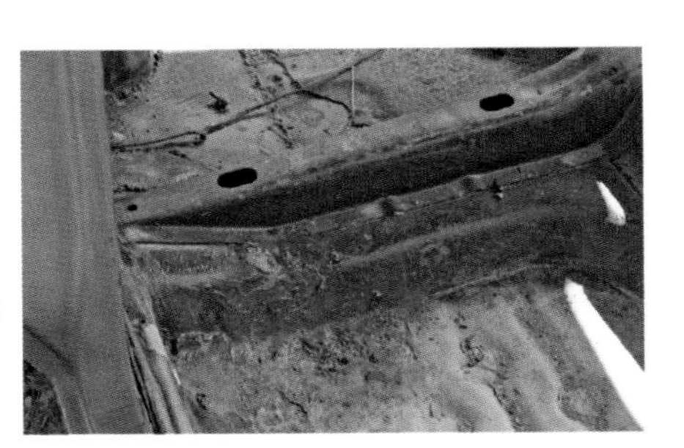

Floor pans rot, so try and lift the carpets to check for this.

wet then expect the worst. Check that the seats are securely attached and look for any signs of welding around the chassis number stamped in the floor (Mk3 cars).

Chassis sections & the underside

Most of the underside will be visible with the car jacked up and viewed from the outside. If you are going to crawl underneath, make sure it's properly supported on axle stands on level, solid ground. Rock the car to double-check that it's completely stable before venturing underneath.

The entire area around the front spring mount must be sound.

The chassis at the rear hanger must be checked carefully, too.

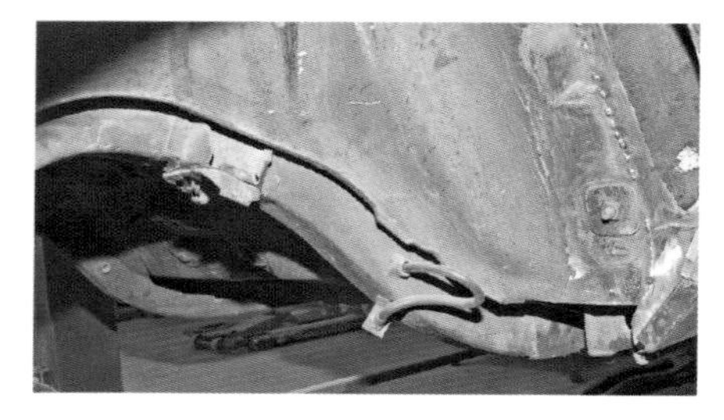

The chassis kicks up at the back arch and corrodes badly here.

Petrol tanks split at the welded seam.

Left: The outriggers and the metal around them are frequently in a bad state.

The most critical section of the underbody is at the back. Follow the rear springs to where they meet their mounts and check carefully for corrosion or previous poor repairs. Make sure that the metal around the shock absorber tops is sound, and

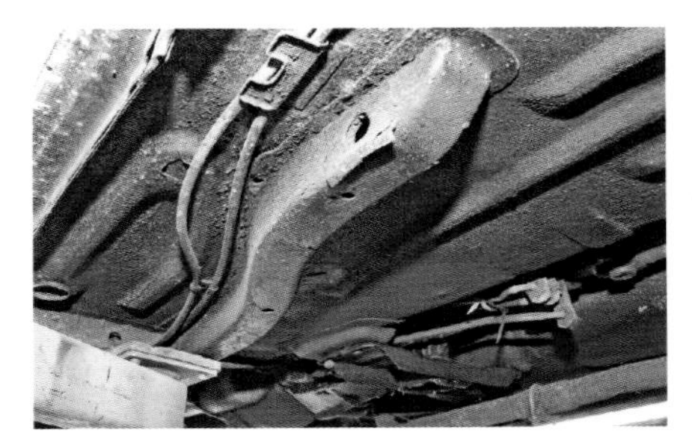

The box section under the floor is vulnerable so check both sides. Make sure that any repairs have been done correctly.

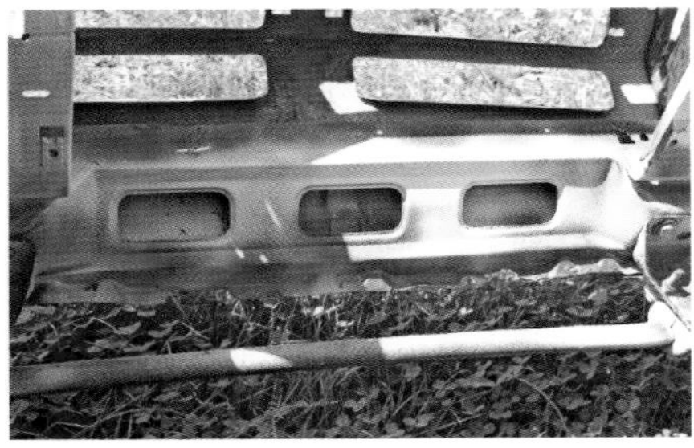

The box section under the radiator is also vulnerable.

then inspect the box section that kicks up over the back axle following the line of the inner wheelarch. This last section may entail crawling underneath a little further as the inside upright of the panel needs checking too. Do not rush this part of the evaluation as repairs to these areas are tricky, time-consuming and expensive. Check the condition of the fuel tank, looking for signs of leakage around the filler neck join or the welded seam around the top.

Moving forward, ensure that the chassis sections are sound; have a good look at the inner sill as you go, especially the join to the floor pans which should also be free of rot or shoddy patches. At the front check the cross member where the anti-roll bar is bolted on, and then the back of the wheelarch/floor join, including the jacking points behind.

Doors rot at the back edge.

The rot starts on the inside.

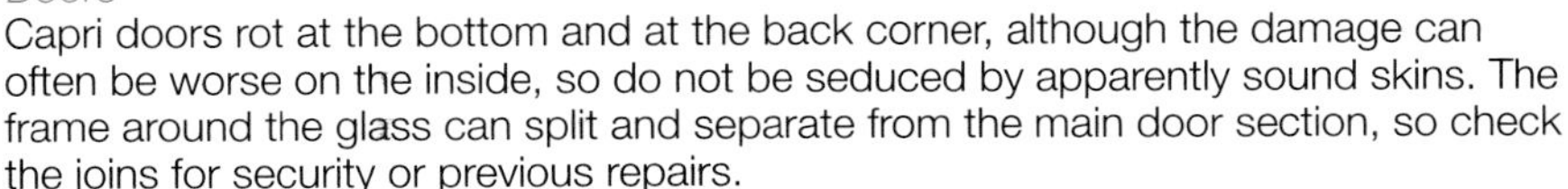

Doors 4 3 2 1

Capri doors rot at the bottom and at the back corner, although the damage can often be worse on the inside, so do not be seduced by apparently sound skins. The frame around the glass can split and separate from the main door section, so check the joins for security or previous repairs.

Rear wings 4 3 2 1

These can be a bit of a disaster area on Capris. The front lower section immediately behind the door rots out above the sill join. The main arch section above the wheel loses its outer skin, and is often filled as a result, but the inner can be even worse.

Even the petrol flap is vulnerable.

The whole of the back arch corrodes badly.

Shine your torch up into the arch to check, then prod (carefully, as there may well be sharp edges due to rusty metal) to check for any problems. Once again, expect to find cheap and nasty repairs, including the use of fibreglass. Have a good look at the lip where the outer and inner arch join as it's very hard to bodge and can give a clue to just how much disguise work has been carried out. The lower section behind the rear wheels also disappears, attacked by road debris on the outside and water leaking in from above on the inside. The fuel filler flap area is also a breeding ground for corrosion and the area under the rear side windows can suffer, too.

The back panel rots around the lights and the boot lid suffers, too.

The back of the roof goes, which is usually more apparent with the hatch open.

Boot & hatch

4 3 2 1

Hatches and boot lids rust along the bottom edge and, like the bonnet, are often worse on the inside. Open the tailgate fully and wait a few seconds before popping your head inside, as weak struts are common and you don't want it dropping on you as you check the interior. The back of the roof will be more visible now and, once again, expect

The spare wheel well collects water – with predictable results.

The rear inner arch to floor join is another suspect area.

blistering or holing along the entire length but especially around the hinge mounting holes. This is tricky to sort out properly so take your time checking.

Hatch seals leak, allowing water to settle in the spare wheel well so take out the tyre and check for puddles and rust. The edges of the boot in the back corners can rust badly, too. Check the whole of the inner wing moulding for rot, especially around the shock mount, plus possible holing at the floor join.

Back panel/rear valence 4 3 2 1

Rust can be found around the rear light apertures and sometimes even under the number plate, starting at its mounting holes. The rear valence splits at the seams and rots on the bottom edge.

The rear valence rots like the front one.

Roof 4 3 2 1

Check the gutter to roof join for corrosion. If the car has a vinyl top make sure that it's still firmly attached, particularly at the front. Run your hands over it all checking for loose patches or unusual lumps which would indicate problems underneath. Any cuts, marks or partial fading can only be properly rectified by replacing the whole lot. If there is a sunroof, make sure that it opens and shuts properly, then check the aperture closely as rot is common in the corners and, although out of sight, a proper repair will still take time. In really bad cases the edge of the roof can be damaged.

Gutter rot is possible, too.

Exterior trim   2 

There really isn't that much of it, but make sure all the badges are present. If the car was supplied from the factory with graphics, check them for damage as replacements may have to be made up specially. Early cars had chromed handles which were made from a cheap alloy that bubbles up over time; nothing can be done apart from replace them.

Beware of vinyl roofs splitting or lifting; they're tricky to repair.

Make sure that all of the locks work and the barrels have not been damaged as Capris were a favourite target for thieves.

4 3 2 1

Wipers

Check that the wipers do not have excessive slop at the spindles; check, too, that the arms are properly attached to them and the blades are not split.

4 3 2 1

Glass/rubbers

The screen should be free of damage or scratch marks in the swept area, or it will be a roadworthiness test failure in many countries. The screen rubber can perish, as can the chromed plastic insert (not fitted on all models) that yellows and becomes brittle over time. Make sure that the door window glass winds up and down to its full extent, and examine the scraper seals which are often split, allowing water into the door cavity. Winder knobs frequently drop off.

4 3 2 1

Lights

Check the headlights for damage to the glass, and for internal condensation which may have discoloured or damaged the silvering. Push them in and out to make sure that the mounting points are secure, as they break easily, especially the rectangular versions. Rear light lenses are harder to find for early cars, so check that they – and the surrounding trim is undamaged. Test the function of all the lights as the rear ones can develop earth faults on later cars. Turn on the side lights, press on the brake, and operate the indicator simultaneously; all three functions should remain clearly separate and unaffected by the workings of the others.

Headlights suffer from condensation and associated problems.

Wheels & tyres

4 3 2 1

Steel wheels are not a problem as parking bumps can be knocked out: alloys, on the other hand, are unable to tolerate such damage so check them carefully, not only at the rim but also in the centre for signs of cracking. Tatty alloys can be refurbished but it's not cheap to do so. Some special editions had wheels which are almost impossible to source now. Check tyres for tread depth and signs of uneven wear, especially on the front due to tracking problems, worn control arm bushes or steering wear. Inspect the spare to make sure that it's legal, correctly inflated and of the same size and speed rating as the others. See if the jack is still there and, if you are really lucky, maybe even the original tool kit.

Bumpers

4   

Early cars had chrome bumpers which suffer from the usual pitting and discoloration associated with that material. Later cars had black painted blades which rust but

Plastic end caps fade, plus become scraped and scuffed.

The area around the front bumper mounts rusts, which also includes the anti-roll bar mounting plates.

are easy to refinish with plastic end caps which should be checked for fading, scrapes, and scoring, and that they're securely mounted. Check the bumper mounts to the chassis on all models, as they also rust.

Interior

Seats

4 3 2 1

The seat coverings are usually cloth, vinyl, or a mixture of both, although leather was used on the Brooklands 280 and the 2.8 Injection Special. The first two materials fade and split, readily and the side bolsters seem to suffer particularly badly on Capris with bucket-shaped seats. The bases sag and collapse and sometimes the frames even crack, as there is a tendency to drop in to the car as it's quite low. Adjustment handles and knobs fall off, so make sure that they're all there.

Make sure that the back surfaces of the rear seats on hatch back cars are not ripped and damaged from sharp-edged loads. Repair to any of the seats can be a problem as covers are not generally available, so cars with tatty ones should be avoided.

Seats fade, split and collapse, so check them carefully.

Lining 4 3 2 1

Head linings split and discolour and suitable replacements are few and far between. Fitting one is probably a job best left to the specialists.

Door trims 4 3 2 1

As with all mass-produced cars, these are a combination of pressed board and vinyl. Damp is the biggest problem, as it causes warping. Holes left by previous attempts at in-car entertainment are a pain, too.

Door cards warp and split.

Mats/carpets 4 3 2 1

Damp and wear cause the demise of most floor coverings; although replacements are available, they do seem expensive compared to some other classics.

Dash 4 3 2 1

The top of the dash cracks, which is not repairable and, in the absence of new parts, a good second-hand one is the only option. Unsurprisingly, these are becoming scarce and are priced to match. Repair skins to cover the damage are available for left-hand drive cars.

Instruments 4 3 2 1

The top of the range cars have a 6-dial dash which includes a large, very legible rev counter. Lower specification models make do with a twin dial set-up which is pretty basic. Most of the instrumentation is reliable but fuel tank sender units become faulty, so check that the gauge works: it may just be the connection at the tank, but the whole lot has to be dropped out of the car for repair. Temperature gauge senders short out internally too but they're cheap and easy to fix.

Controls 4 3 2 1

Heater blower motors fail, so check that it operates on all settings. Switches lose their markings and replacements are hard to find for earlier cars if any of them are malfunctioning, so check that they all do their job. Stalks can sometimes break – the indicator one most often – a defect which will be obvious. Check that the gear lever knob is secure as it can pull free from the threaded insert which screws on to the lever.

Steering wheel & column 4 3 2 1

Padded wheels can split, and the vinyl becomes threadbare and sticky with age, neither of which is a tactile pleasure! Column bushes wear, so grab the wheel top and bottom and push and pull it up

Instruments are pretty reliable in the main, but still check that they work.

and down; it should not be possible to feel any looseness or play. Check the plastic column shrouds for cracks or splits from theft attempts, and make sure that the steering column lock works properly. It's also a good idea whilst there to check the ignition key, as it can fracture, leaving the remains stuck in the barrel.

Mechanicals

All engines 4 3 2 1

The Capri was fitted with a wide range of engines, and a few of their individual foibles are looked at later, though some checks are applicable to all. Most will leak oil from somewhere, but make sure that it's not too heavy. Be suspicious of something that is too dry as it's likely to have been steam cleaned ready for sale. Check for water leaks, or staining from head joints, water pumps and pipe connections.

1300 OHV engines are rare but reliable.

Kent 1300 & 1600 4 3 2 1

These motors are pretty strong though high mileages result in piston ring/bore wear, so remove the oil filler cap and watch for fuming. Timing chains rattle, and worn cam followers can tap noisily. Once very common, this engine is now getting harder to find second-hand, and some parts needed for rebuilding may take some tracking down.

2-litre V4s are not regarded as Ford's best units.

V4 4 3 2 1

Not one of Ford's best designs, but the German-built version is much more reliable than the UK manufactured one. Fibre timing wheels break up and, although replacement steel versions are available, they're not cheap and increase engine noise. Oil pump drives can fail without warning, with catastrophic results. Inlet manifolds can warp, leading to air leaks, associated poor running and overheating. The internal balance shaft wears out its bearings, resulting in excessive vibration and noise. Camshaft bearings can spin in their housings, cutting off oil supply. Parts are a problem generally but less so for the Cologne-built motor as it ended up in many continental Fords, as well as some Saabs.

Pintos are pretty robust if looked after.

Pinto 1600 & 2000 4 3 2 1

The main problem with the Pinto is an

inadequate oil supply to the overhead camshaft, due to a spray bar that isn't up to the job, a problem compounded by owner neglect so that old, sludgy oil blocks the jets. Avoid any which obviously rattle from the top end, or budget for head removal for a new cam and followers. The bar is very cheap to buy, so treat it as a service item and change it every three years. Expect piston ring/bore wear once past 80,000 miles, otherwise it's a reliable engine, as long as the cam belt is changed regularly. Parts are easier to source than for earlier engines.

V6 Essex & Cologne

4 3 2 1

Once again, the German version is stronger, and arguably more reliable. Both can suffer from overheating problems, as internal rusting of the cast iron block can cause silting in the narrow passages in the radiator, and overstress the already marginal cooling system. Timing gears break, as on the V4s, as does the oil pump drive, and the problem with cam bearings remains. Parts availability for both versions is pretty good.

V6s provide the best performance and are the most sought-after engine.

Gearbox

4 3 2 1

4-speed boxes are robust and a pleasure to use with a fast, smooth gear change. High mileages will result in general bearing wear with weakening synchromesh usually affecting second gear the most.
5-speeders are not regarded as quite so reliable, and can lose top gear if they're run low on oil for any length of time. 2nd is likely to show signs of wear first, just like the earlier box. Make sure that there is no sign of play in the rear bearing on either type: check by pushing the prop flange up and down and looking for movement; look for deterioration in the rubber crossmember bush as well. Gear levers become sloppy as bushes wear.

Check the rubber bush in the gearbox mount.

Automatic box

Early cars had a Borg Warner box available as an option, but it was never popular, and therefore very rare today. Mk3 cars had the Granada-derived C3 box which needs regular maintenance to ensure reliability, so check the service records to make sure it has received it.

Prop centre bearings need checking.

The back axle end plates become porous thanks to rust.

Back axle/prop shaft

4 3 2 1

Two types of rear axle were fitted to Capris, the differential is robust in both, but will obviously suffer if the oil level has been allowed to drop, so look for signs of leaks at the nose, and at the join to the rear brake back plates. The steel end cover, where the level plug is, rots too and becomes porous. 2.8 Injections and 280s with a limited slip diff use special oil, so check that the owner has been using it or damage may have been caused. The prop centre bearing can wear allowing it to drop, so give it a wiggle to check. Look for signs of wear in the universal joints, too.

Radiator & cooling system

4 3 2 1

The radiator is readily accessible on the Capri and it's fairly easy to spot leaks, repairs or damage. If the car is fitted with a viscous fan, make sure that it's free

Cooling systems are marginal in larger engined cars, so check for leaks and signs of overheating.

Viscous fans seize up.

to spin, as they regularly seize and replacements are not cheap. Check the hoses for signs of cracking. If an expansion bottle is fitted, make sure that it's still clear and not discoloured by rusty or oily scum which would indicate head problems. If you suspect there is a problem leave the engine idling as you continue your inspection, and return once it's warm and squeeze the hoses, if the cooling system is pressurising they will be rock hard to the touch. At the same time make sure that the thermostat is functioning properly, and the coolant is circulating as it should. If you think there is a question mark over circulation, turn off the engine and check that the radiator is hot all over, but be careful not to burn your hands! These checks are important for all models, but essential for V6-equipped cars where the cooling system may be struggling.

Brakes

4 3 2 1

Front discs should be checked for scoring, wear and rust. They're also prone to warping which will show up during the test drive. Calipers seize thanks to internal corrosion from unchanged brake fluid and from damaged dust seals; replacement is straightforward and they're fairly cheap. Check solid brake lines for rust throughout the car, although the ones at the back are the most vulnerable. Flexible hoses should be examined for splitting or other signs of deterioration, or even contact with other components. Brake back plates should show no sign of leaks from either the top, around the cylinder mounts, or at the bottom, which would indicate fluid leaking internally.

Check brake pipes for corrosion.

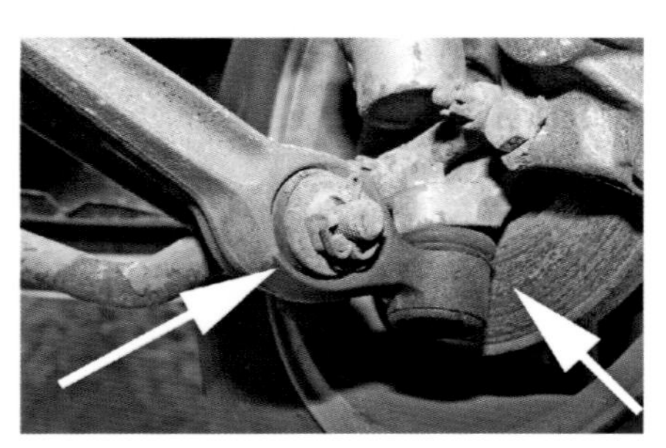

The anti-roll bar bushes and the bottom ball joints are prime candidates for wear.

There is a lot to check in the hub area.

Hubs, steering joints & rack

4 3 2 1

With the car jacked up grasp a wheel at the top and bottom (6 o'clock position) and pull it in and out. Any play will be either the ball joints or wheel bearings. Get someone to apply the brake whilst you repeat the exercise: if the play has gone then it's the bearings, which will need adjustment or replacing. Try spinning the wheel to see if you can hear them grinding, rotating it very slowly by your fingertips may reveal roughness, too. If the play is still there then it must be caused by wear in the bottom ball joint.

Continue rocking the wheel and have a look for play in the top of the suspension strut; a small amount is inevitable, but anything more means new bushes. Move your hands to the sides (9.15 position) and wiggle once again; movement felt now will be in the track rod ends or the rack, with the former being the most likely

The steering coupling in the engine bay splits.

candidates. Pull the wheel on to full lock and have a look as you move the rim, watch each in turn until you have located the play. Check the rack for split boots or leaks and make sure the mounting rubbers are not perished. There is a rubber coupling in the steering joints, visible in the engine bay, which should not show signs of tearing, sagging or other wear.

If the car has power steering, check the condition of the fluid – which should not look black or be gritty – and ensure that the level is correct. The wear tests are the same as with a manual rack, also make sure that all the pipework is secure and that there are no leaks. On both types of rack, move the rim slowly from lock to lock so that the full extent of steering movement is checked for stiffness or sections that are looser than the rest.

The track control arm bushes wear readily on Capris, too, so check them for wear by pushing and pulling the rim. They're often oil-soaked and swollen, which will be immediately obvious. Look at the anti-roll bar bushes in the middle of the arm, too, as they have a very short lifespan. Follow the bar up to the chassis and check the rubber mounts there as well.

Battery & charging 4 3 2 1

No real weak spots here; listen for noisy alternator bearings when the car is idling, and wobble the pulley when the engine is off to detect wear in the bearings, a test which applies to early dynamo-equipped cars, too. Take a multimeter with you if you want to fully check the state of the charging circuit. A good battery will read 13 to 13.5 volts at rest, start the engine and this should rise to 14 to 14.5. Check that it does not exceed this figure and that the lower figure is achievable when there is a load by turning on the headlights, indicators, heater fan, etc. If a dynamo is fitted, the charging rates, although similar, may take longer to be achieved and at higher revs.

Electrics 4 3 2 1

The fuse box can cause all sorts of electrical gremlins due to old and insecure connections. The rear wiper often appears faulty, but it may only be the exposed connections in the tailgate. The wiring in general is no more or less reliable than other cars of the time, but vehicles of this age will often have acquired DIY additions or stray circuits added to bypass problems, so if you spot any ask the owner why it has been done.

Lots of electrical gremlins can be traced back to poor connections in the fuse box.

Carburettor

Ford Motorcraft 4 3 2 1

These are basic, reliable carburettors but not particularly long lived. Spindle wear is a problem and many have been replaced by Webers over the years.

Auto chokes are unreliable.

Weber 4 3 2 1

These are petty robust units which also suffer from spindle wear after a high mileage, so check by trying to wiggle the linkage in relation to the body.

Ford VV 

Universally disliked, these carbs suffer from weak diaphragms and a hopeless automatic choke. Many will have been ditched completely by now, or, at the very least, have a manual choke conversion fitted.

Injection   

The system itself is pretty reliable but is let down by defects in both the wiring and the sensors that provide the information. Fault-finding can therefore be a tricky business so haggle hard if the car has a running fault.

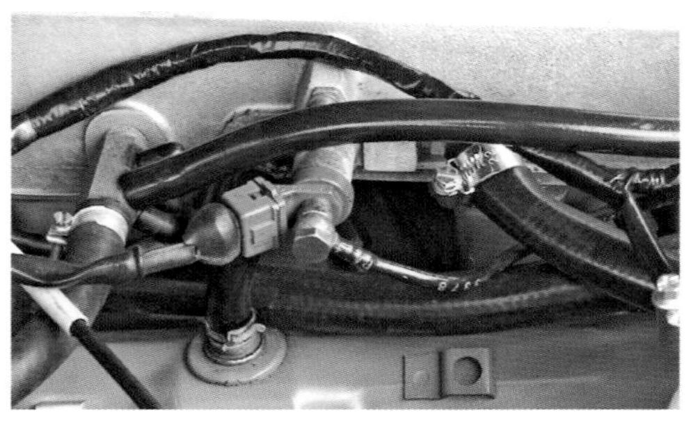

The fuel injection system is sometimes let down by poor connections and sender units.

Exhaust   

The system is completely conventional, so its condition is more likely to be a haggling point, rather then an item to put you off the car, especially with V6s where a complete replacement is not cheap. If you can hear the distinctive chuffing of a leak under the bonnet of a Pinto-equipped car, it may be the exhaust manifold gaskets, which are pretty shortlived, or it may even be a cracked manifold, as they can split between the cast curves.

Pinto exhaust manifolds can split.

Shock absorbers    

Once again a conventional set-up, so check for leaks or damaged bushes, then carry out a bounce test: one and a half oscillations before the car comes to rest is fine, any more and the units need replacing, and always in matched pairs. On the mechanical side of the suspension, check the front coil springs, as these can sometimes snap at the ends.

The axle locating plates sometimes work loose; check for signs of movement.

The rear spring eye bushes need checking.

Rear springs 4 3 2 1

Check the leaves for cracks or signs of the straps splitting or falling off. Severe spring eye bush wear may be obvious; lesser amounts will mean using a lever to pull down the assembly in relation to the pin, and looking for play. Check the shackles and axle mounting bolts for signs of corrosion or movement as the latter can loosen off with time and wear.

Test drive 4 3 2 1

This should be at least 15 minutes long, and include as many different road types as possible. It's essential to escape purely suburban roads to get a true feeling of how a Capri drives. Make sure you are happy with the seat position, mirror adjustment, and layout of all the controls before you set off as your thoughts must be concentrated on how the car performs, not on fumbling around trying to find a particular button.

Starting/warning lights 4 3 2 1

All engines should fire promptly without long-winded churning on the starter, and settle to a smooth idle immediately. The oil light should extinguish immediately and stay out. If an oil pressure gauge is fitted, it should climb instantly once running and should never drop towards zero, even when the engine is hot and idling. The coolant temperature needle should remain in the middle, although Fords do have generous latitude here, so it's probably okay as long as it's not nudging the red. Some versions will also have a voltmeter fitted, so make sure that it registers correctly.

Clutch 4 3 2 1

The pedal should feel firm and bite half-way up its travel. Too high and the clutch is worn, too low and it needs adjustment. As drive is taken up it should be smooth; any judder would indicate oil contamination. Gear selection should be easy, if a little notchy into first.

Gearbox

Listen for whining from each ratio and try and catch out the synchro with some fast down changes. Accelerate and decelerate sharply in each ratio to see if the box wants to pop out of gear. Changes, in general, should be clean and easy.

Auto box

4 3 2 1

Regardless of the type fitted, drive selection should not be accompanied by a clunk. Changes on the move should be smooth, and the kickdown should provide an immediate response when the throttle pedal is floored.

Differential

4 3 2 1

These are pretty strong but listen for a whine from the back end. Rear wheel bearing wear can sometimes be confused for diff problems, but their noise should rise and fall as the car moves to turn left or right, whereas the diff will remain pretty constant. Vibration may be felt from a worn propshaft which manifests itself by a low juddering felt through the floor.

Steering & suspension on the move

The steering should be light above parking speeds and free from any knocking or vibration. Worn track rod ends, or track control bushes, can impart a woolly feeling to the whole process; any shuddering vagueness over bumps will almost certainly be caused by the latter. A Capri should feel taught, though tempered by enough refinement to prevent any unpleasant jarring.

Braking performance including handbrake

Before driving off, it's worth checking the state of the brake pedal which should be firm under foot. Apply gentle pressure and try and push it down very slowly; if it sinks under this soft but constant load then the seals in the master cylinder are shot. This can be double-checked by taking your foot off and quickly reapplying it, which will immediately give a firm pedal again. If the car has a servo fitted, pump the pedal and hold the pressure on, then start the engine. The pedal should drop half an inch or so instantly if the servo is working correctly. On the move, the brakes should pull up the car in a straight line; if there's a bias to one side a caliper may be sticking. If the discs are warped you will feel a pulsing sensation through the pedal as it's applied.

Servo and master cylinder operation should be checked before venturing out on the road.

The back brakes will have a

self-adjusting mechanism, which almost certainly won't be working properly, so the handbrake may have long travel; it should pull the brake on fully in 4 or 5 clicks. Try pulling away from rest with it applied; the car should squat down evenly at the back as you do and provide good resistance. This test can also reveal a slipping clutch, although the car's owner may not be too impressed with the process.

Final thoughts 4 3 2 1

Once back after the test drive leave the engine running and walk around the car once more to double-check that nothing is leaking, or that the engine has not started to smoke after its work-out. Pop open the bonnet to double-check under there, too, before retiring to consider the overall verdict.

Evaluation procedure

Add up the total points. Score: 236 = excellent; 177 = good; 118 = average; 59 = poor.

Cars scoring over 165 will be completely useable and will require only maintenance and care to preserve condition. Cars scoring between 59 and 120 will require serious restoration (at much the same cost regardless of score). Cars scoring between 121 and 164 will require very careful assessment of necessary repair/restoration costs in order to arrive at a realistic value.

10 Auctions
– sold! Another way to buy your dream

Although mainstream Capris are not yet a regular sight at classic car auctions, their growing appeal and rising values will undoubtedly see that situation change over time, so it's worth taking a moment to consider another route to your dream purchase.

Auction pros & cons

Pros

Prices tend to be lower than those asked by either dealers or private sellers, so it could be the chance to pick up a bargain. The auctioneers have usually checked out the legal status of the car, and relevant paperwork can often be viewed before the sale.

Cons

There is often little opportunity to inspect the cars fully, and they cannot be driven. Many cars end up in auctions as they require some work, the trick is to spot exactly what that may be and factor it into your bids. It's very easy to get caught in an upward spiral of bidding and spend too much. There may also be a buyer's premium to add to the hammer price.

Which auction?

Classic/collector car magazines have listings of forthcoming auctions, and most auctioneers have a website detailing the vehicles, usually accompanied by a photograph. There may be details of previous sales and the amounts reached by each lot, which will give you an idea of the state of the market.

Catalogue, entry fee & payment details

Entry to the sale and any viewing days is often included in the purchase price of the catalogue. This document will also list each lot with a brief description of the car, any paperwork, such as service history, MoT etc, and mileage. It may also include a guide price which the auctioneers expect it to make, and any premiums due if sold. Once the hammer has fallen, payment of a deposit is usually expected immediately, with any outstanding balance paid within 24 hours. Read the catalogue carefully, especially about payment methods as auction houses vary in what they will accept, and may charge extra for the use of your flexible friend. Whatever method you choose, the car will not be released until payment has cleared, and if you delay in sorting it all out, a storage charge may be added.

Buyer's premium

A buyer's premium will almost certainly be added to the final hammer price; remember to factor it into your bidding. There may be a state or other local tax on top as well.

Viewing

It's likely that there will be a set viewing time either immediately before the sale, or preferably a couple of days in advance. Ask for the bonnet and boot to be opened and the engine started, plus check out all available paperwork for the car that interests you. It may be possible to have the car jacked up to check suspension, steering, etc, so take advantage of any opportunity to examine the vehicle as closely as possible.

Bidding

Auctions are exciting, especially when several people are after the same car, and it's all too easy to get swept along in the euphoria of it all. So, before you start down this potentially expensive road, decide on the maximum amount you're willing to pay and stick to it. If you're unfamiliar with auction procedure get there early, register to bid, then settle back and watch how other people do it. When your car arrives, bid early so the auctioneer knows where you are, he/she will keep coming back to see your reaction to other bids. If it's time to duck out make it very clear, a vigorous shake of the head should be enough. If you win, your card number will be taken and you will have to go and pay your deposit. If the car you want fails to sell as it has not reached its reserve, it may still be worth going to the office and registering your interest, as the auctioneers may contact the owner with your offer to see if they wish to sell at that price.

Successful bid

With the car secured and paid for, the next issue is getting it home. If you cannot drive the vehicle (no insurance or MoT, for example), either hire a trailer or arrange for one of the many transport firms to shift it for you; the auction office should have a list of suitable companies. Some sales also offer immediate short term insurance cover but this is rarely the most cost-effective way of dealing with the problem; a phone call to your usual broker will almost certainly result in something cheaper.

Internet auctions

There are undoubtedly bargains to be had via internet auction sites, but the overriding rule is never to buy anything unless you've seen it in the metal, or it's so cheap that any spares gleaned from the car would be worth more than you paid. As usual, distance comes into play here, so try and limit your search to an area close to home. Once again, it's possible to get carried away so set a limit and don't be tempted beyond it in the last few minutes of the auction when things hot up. It's very hard to get your money back if you are misled and, unfortunately, there are also cases of outright fraud, so be careful.

Auctioneers

Barrett-Jackson	www.barrett-jackson.com
Bonhams	www.bonhams.com
British Car Auctions	www.british-car-auctions.co.uk
Cheffins	www.cheffins.co.uk
Christies	www.christies.com
Coys	www.coys.co.uk
H&H	www,classic-auctions.co.uk
RM	www.rmauctions.com
Shannons	www.shannons.com.au
Silver	www.silverauctions.com
eBay	www.ebay.com/www.ebay.co.uk

11 Paperwork

– correct documentation is essential!

The paper trail

With the Capri now firmly in the classic and collector sector, the amount of paperwork available to document the history of any prospective purchase is very important. Old receipts for work done, particularly from known specialists, will make the vehicle more attractive to future buyers, so never throw anything away.

Registration documents

Virtually every country has a system of recording vehicle and owner details on a single form – the log book or V5C in the UK, Pink Slip in the USA, for example. Check that all the information on it is correct and that it relates to the person selling the car and the car you're actually looking at. If the vehicle is registered overseas complications may arise, but, within the EEC, things are a lot easier than they used to be. Check with your relevant licensing authority beforehand, though, as registration may be a long-winded and potentially expensive process. In some countries a record of previous owners can be obtained (after paying a small fee) which often throws up interesting results.

Roadworthiness certificate

Wherever you live there is likely to be a regular test to check the roadworthiness of your car and the amount of pollution it's creating. In the UK a certificate is issued when the vehicle passes the annual test, permitting its use for 12 months. A valid certificate is essential with your purchase unless you are specifically buying a restoration project. Old certificates are often kept by owners, and these show accumulated mileage and add to the general history of the car.

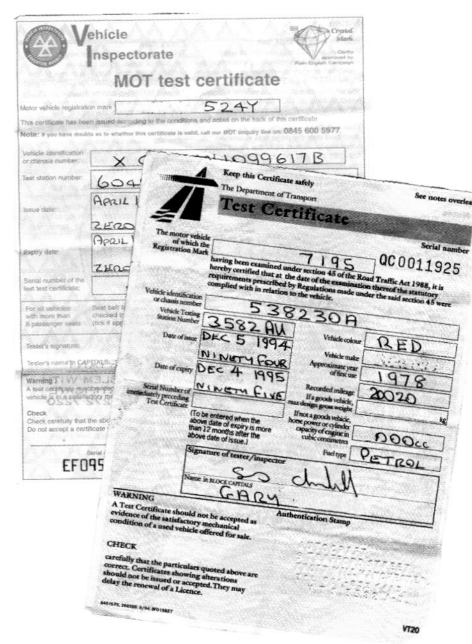

Vehicle Inspectorate

MOT test certificate

Motor vehicle registration mark: 524Y

Note: 0845 600 5977

Keep this Certificate safely

The Department of Transport

See notes overleaf

Test Certificate

The motor vehicle of which the Registration Mark: 7195

Serial number: QC0011925

having been examined under section 45 of the Road Traffic Act 1988, it is hereby certified that at the date of the examination thereof the statutory requirements prescribed by Regulations made under the said section 45 were complied with in relation to the vehicle.

Vehicle identification or chassis number	538230A		
Vehicle Testing Station Number	3582AU		
Date of issue	DEC 5 1994	Vehicle colour	RED
	NINETY FOUR	Vehicle make	
Date of expiry	DEC 4 1995	Approximate year of first use	1978
	NINETY FIVE	Recorded mileage	20020
Serial Number of immediately preceding Test Certificate		If a goods vehicle, max design gross weight	kg
(To be entered when the above date of expiry is more than 12 months after the above date of issue.)		If not a goods vehicle, horse power or cylinder capacity of engine in cubic centimetres	2000cc
		Fuel type	PETROL

Signature of tester/inspector

Name in BLOCK CAPITALS: GARY

Authentication Stamp

WARNING

A Test Certificate should not be accepted as evidence of the satisfactory mechanical condition of a used vehicle offered for sale.

CHECK

carefully that the particulars quoted above are correct. Certificates showing alterations should not be issued or accepted. They may delay the renewal of a Licence.

VT20

Even old test certificates can increase the value of a car.

Road fund licence

Virtually every country has a form of tax that must be paid in order to use your vehicle on the road. There's often a sticker or disc that must be displayed in the windscreen to prove compliance and the dates of validity. If the proposed purchase isn't taxed make sure that any other relevant laws have been adhered to. For example, in the UK a SORN declaration (Statutory

Off Road Notification) must be made or the owner is fined. If the car has been laid up for a very long time, it may not have the latest registration document and so have avoided this paperwork, but as the new owner you will have to sort it out.

Valuation certificate

If the car in question is insured under a classic car policy, the owner may have had to arrange an independent inspection and valuation to satisfy the insurers. Ask to see it, and for a copy if you buy the car. Do not take the opinions expressed in the document as fact, as some specialists can be lenient in appraisals for insurance purposes, though it should help to confirm some of the points you've spotted whilst working your way through the inspection laid out in this book. If there's no certificate and you need one, the first port of call is the owner's club, which should have an officer capable of providing the relevant information, and putting you in touch with someone able to inspect the car and issue a certificate.

Service history

It's unlikely that you will find a Capri with a full dealer service history, but records of regular care from a specialist are just as valuable, and even those from a local garage are better than nothing. Many cars are maintained by their owners, so ask how often the oil was changed for example (and every 3000 miles is what you want to hear), and the quality and specification of the lubricant used. Even the brand of oil filter chosen can give an indication of the amount of care taken during servicing. A knowledgeable seller should be able to answer servicing questions promptly and in some detail. Any receipts for parts or work are useful and should be retained.

Restoration photographs

Most people take photos during a restoration for their own pleasure, but also know it will ultimately enhance the value of the car when it's time to sell. Make sure the car in the shots is actually the one you're looking at; compare backgrounds if it's a home restoration, or try and look for areas where fresh work can be readily verified. Make sure that the seller will hand over the photos, or a least copies or scans if you buy the car; they will be just as valuable to you in the future.

12 What's it worth to you?

– let your head rule your heart!

Having worked your way through the inspection chapters you should have a good idea of your proposed car's condition, whether a total basket case but suitable for restoration, or a clean, useable car ready to drive away. If you've done your homework and checked the price guides in classic car magazines, and gone to a few shows or kept an eye on the internet ads for a time, you should be able to decide if that condition merits the price being asked.

The structural or mechanical solidity of the vehicle may not be the only factor to take into account, though, as other issues like rarity also have a bearing. It will always be a better bet spending money bringing a 'Brooklands' up to scratch, than a 2-litre Pinto in the same condition. As a general rule, the newer the car that you are looking at, the higher the score it needs to have accumulated, using the Essential Buyer's Guide points system, to make it worth its money.

Tuning is another area which will have an impact on how attractive a car may be, and historically the Capri has been popular with owners looking to improve and enhance the factory offering. Value will only be increased if the work has been done to a high standard, backed up by appropriate receipts from a recognised specialist in the field. Home tuning may have a negative affect on the car's perceived worth, as will cheap fibreglass add-ons like spoilers and sill kits. Garish, non-standard paint schemes will put off far more potential buyers than they attract, so although bright pink might be your favourite colour it will do little for the resale value. Interior upgrades, or the addition of badges and decals from higher specification models, may make the car more luxurious or appear more attractive, but ultimately will do little to enhance the value as they're not original. This holds particularly true for earlier cars which will be worth most when they're as close as possible to how they rolled off the production line.

Once you have decided how much the car is worth to you, never be afraid to haggle. List some of the defects that you have found as justification for a price reduction, it may not work, but the worst that can happen is that you will end up paying the full asking price. Even if the car is in good condition do not shy away from mentioning the mundane such as worn tyres. Dealers, in particular, are more flexible as they will have built in a fair margin to begin with to cover warranty work or rectification of minor problems pre-sale, and are more likely to chuck in small extras to sweeten the deal.

Small original details can enhance a car's worth; non-original bits may have the opposite effect.

13 Do you really want to restore?

– it'll take longer and cost more than you think

The idea of restoring a car is very tempting on many levels, especially if you do it yourself, whether it's the sense of achievement in having saved a Capri from the scrap man, the acquiring and exercising of new skills, better use of your time rather than hours glued to the idiot's lantern or, last but not least, the chance to save some hard earned money.

Bear in mind, however, that the Capri's relatively low market value is in direct conflict with the high cost of repair panels needed to rescue a rusty car, so it's always better to buy the best car that you can afford. If you do opt to do it yourself, even if it's only a few parts of the car, be ruthless in assessing your abilities: can you successfully chop out rot and replace it with new steel without distorting some vital section? Will your paintwork come up to scratch? How many engines have you successfully rebuilt? The skills needed to sort out these problems can be acquired, but have you enough spare time to do that? Have you got the space, the tools – or enough obliging friends – to lend them to you? Finally, it's very easy to underestimate just how much time is needed to restore a car properly, so have a chat with someone who has done a Capri already. If they're honest they'll likely tell you that it's best to double the amount of time and money you think it will take to see the car completed. This list of potential troubles is not an attempt to put you off, but note that the small ads in magazines are usually well padded with adverts for 'unfinished projects,' and you don't want your dream to end up that way, particularly as it usually entails a financial loss to boot.

Do you have the skills to carry out major restoration work?

The time factor associated with a DIY restoration can be cut down by doing a rolling restoration, but this has its own inherent problems, as jobs do not always conveniently fit into the slots you have allocated for them, and part-restored cars that look a mess tend to attract unwanted attention on today's roads. Paintwork doesn't fit this scenario very well either, so this at least would really have to be done in one go, unless you want to deal with soggy filler, peeling primer and ingrained dirt on every panel.

If you decide to have a professional do the work, the process is potentially just as troublesome, and will most definitely be expensive, as restoration is labour intensive and will almost certainly cost you far more than the market value of the car. If your Capri needs a respray, for example, are you happy with just a flat off, spot prime and top coat, or are you expecting a full glass out, bare metal job? Unless you make your exact requirements crystal clear and commit them to paper, there is always the chance of disappointment. One area where you definitely do not want any surprises is what it's all going to cost, so always get a written quote stating precisely what will be done, for how much and by which date. Try and get something included to the effect that the final amount will not exceed the quotation by more than 10 per cent. Before entering an agreement, check out the firm's reputation and look at some of its previous work, especially the most recent. Speak to other people who have used its services about their experiences, both good and bad. The few established Capri specialists will have a waiting list, so it could be some time before your car reaches the head of the queue.

Last, but certainly not least, compare the price of cars that are already restored against the likely amount you'll have to stump up to get yours done as there may be a financial advantage in simply buying one 'off the shelf,' with the added bonus that you can enjoy the Capri experience straight away.

How far you need to go to repaint a car is a matter of opinion; always discuss the matter fully with your chosen restorer to avoid misunderstandings.

14 Paint problems

– a bad complexion, including dimples, pimples and bubbles

Fading

This afflicts red paint hues more frequently than other shades, but a lack of paint care can see it affect all colours in the end. Either way, the situation may be rescued by T-Cut or a similar compound, although if that fails to restore the shine a repaint is in order. The lacquer used as the top coat on metallic cars can also discolour over time and become slightly milky, which produces a similar effect.

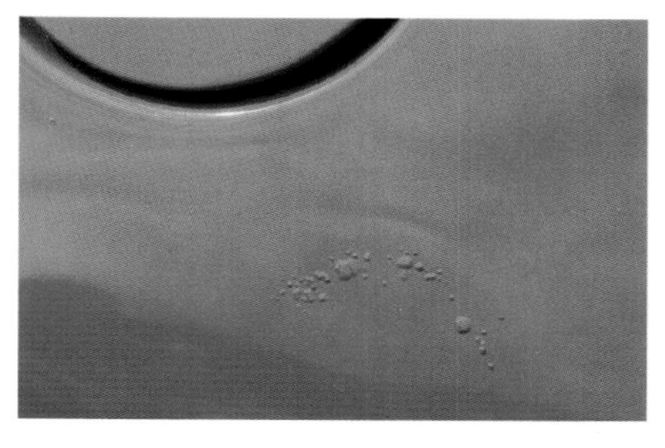

Bubbling paint means corrosion lurks below.

Bubbling

This is a result of rust forming under the paint layer (or filler) and swelling up. The spread of the corrosion is always greater than the visible area, and when the old paint is removed thin threads of rust will be found creeping out from the centre of the problem. Removal of all the old paint from around the affected area and rectification of the rust problem is the only answer.

Crazing is unsightly: this will have to come off and be done again.

Crazing/cracking

These two problems share similar roots: either the surface was inadequately prepared for the new paint, there was a reaction to old material left on the metal, there was some contaminant present before the top coat went on, or the paint was applied over filler which was not cured properly. To make sure that the problem doesn't return, take the surface back down to bare metal before filling and repainting.

Orange peel

This is a result of too much paint being applied during a re-paint, and the surface really does resemble orange peel. This can usually be sorted by flatting off with fine wet and dry paper to level the surface, followed by buffing. This is labour intensive however, so expensive if you are paying someone else to do it.

Orange peel can be sorted out by flatting and buffing.

Micro blistering

These are tiny blisters of solvent or water trapped under the new paint, and as such are not easily treated without stripping it all off and starting again.

Micro blistering is caused by solvents escaping from under the top coat.

Silicone dimples

If the car has had a quick, poorly prepared respray or other paintwork, silicone residues from the workplace can be left under the new paint, which then looks like little puddles in the surface. These can only be rectified by repainting.

Peeling paint/lacquer

If the primer has been inadequately flatted or has a contaminant on the surface, the top coat will not adhere properly and can flake off. If this happens to a clear over base finish, the whole lot will have to be taken back down and re-done. Solid coats may respond to a localised repair.

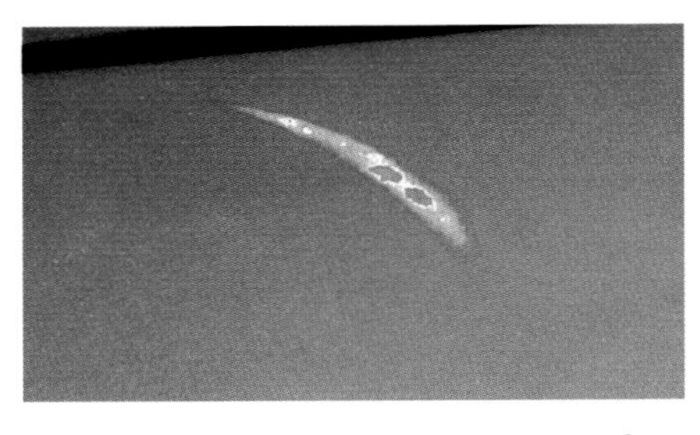

Peeling lacquer is a big problem as the whole panel will need stripping back.

Dents

Small dents can be removed by specialised firms, many of whom offer a mobile service, so check in the *Yellow Pages*. Larger dents will require more attention; given the prices charged by body shops, make sure not to overlook the potential cost of getting your car straight.

15 Problems due to lack of use

– just like their owners, Capris need exercise!

Hydraulic problems

Brake fluid is hydroscopic and attracts water, which promotes internal rusting of metal components. Expect seized wheel cylinders and caliper pistons if the car hasn't been moved for a long time, and even a short period could produce enough contamination to give a spongy pedal. Leaks are common when the car is then put back into use.

The braking system will not tolerate long periods of idleness.

Coolant problems

The cast iron blocks used in Capris rust internally if there is a low concentration of anti-freeze, as it also contains corrosion inhibitors. Flushing will remove light silting; heavier deposits will need chemical treatment to shift. If the car has been left in this state for too long, then the heater and main radiator may be terminally damaged through silting up.

The waterways will sludge up if left short of anti-freeze/inhibitor as the engine block rusts internally.

Electrical problems

Alternators, dynamos, starters, electric fuel pumps and mechanical voltage regulators do not like inactivity. Corrosion builds up on exposed brushes and their tracks. It's fairly common for these items to cause problems shortly after the car is pressed back into service, particularly on the charging side, when even solid state regulators seem to give up the ghost. No battery will stand long term inactivity well, and recharging will fail to revive them if the debris usually held in suspension in the electrolyte has settled and shorted out the plates. Wiring connections oxidise and frequently require separating and cleaning, as do bulb holders.

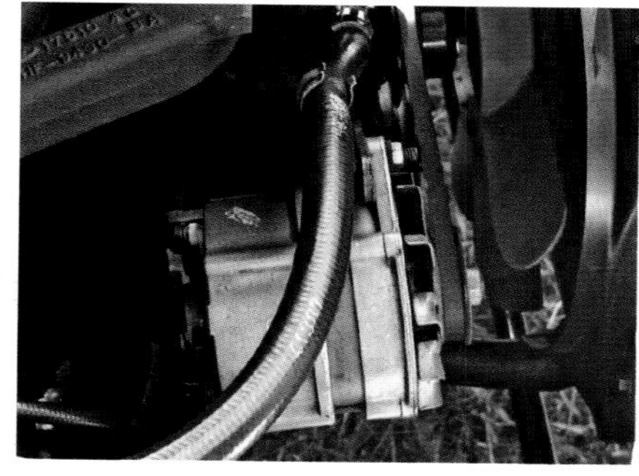

Electrical items often fail a short time after being put back into use.

Oil contamination

Engine oil picks up several nasty contaminants when in use, including acids, and if the car has been used regularly for short trips, water from condensation can be a problem. When the vehicle is then stood idle, these promote rusting and deterioration in bearing surfaces and on cylinder walls. Change the oil as soon as the car can be run, and then again a short time after the vehicle has been put back on the road.

Tyres

Tyres have a finite life of around five years, by which time they'll have started to harden and deteriorate. If left sitting unused, and particularly if under-inflated, there is a real chance that they will go out of shape and retain that deformity for good. Check the side walls carefully for cracking and date of manufacture (often an obvious code)

Exhausts quietly rust away from the inside out.

Exhaust

All petrol cars produce condensation until they have reached operating temperature, which collects in the exhaust and rots it from the inside out. Cars left to stand usually have some moisture in the system and acquire more from the atmosphere, so expect anything that has been sitting for any length of time to have a well corroded exhaust, regardless of how good it may look from the outside.

Unleaded fuel 'goes off' leaving behind a gummy deposit.

Stale fuel/pump problems

Modern unleaded fuel goes stale more readily than the old leaded variety, leaving a sticky deposit when it evaporates. This clogs jets and clings to the float bowl where it becomes a hard, semi-fossilised coating if left too long. Any car which has been unused for more than 12 months would probably benefit from a carburettor strip and clean. No fuel pump likes standing for extended periods either, and sometimes will need back filling to prime before it functioning properly again, although the diaphragm (mechanical type) may be beyond saving if it has been allowed to dry out completely.

16 The Community

– key people, organisations and companies in the Capri world

Due to ever changing contact details particularly, of club officials, the listings below have been restricted to web and email addresses which should remain more constant. Land line and postal addresses are available from the sites.

UK clubs

Capri Club International
www.capriclub.co.uk
Email capriclub@btclick.com

Capri Club Scotland
www.capriclubscotland.co.uk
Email capriscotland@aol.com

Capri Mk1 Register
www.fordcapriclub.com
Email information@fordcapriclub.com

Caprisport
www.caprisport.com
Email caprisport@tesco.net

Continental clubs

Capri Club Denmark
www.capriclub.dk
Email info@capriclub.dk

Capri Club Deutschland
www.capri-club-deutschland.de
Email info@capri-club-deutschland.de

Capri Club Italia
www.capriclubitalia.it
Email info@capriclubitalia.ot

Capri Club of Ireland
www.capriclubireland.com
Email capriclubireland@yahoo.com

Capri Club Passion
www.ford-capri.net
Email Capri.Passion@ford-capri.net

Capri Drivers
www.capridrivers.be
Email info@capridrivers.be

Capri Register Austria
www.capri-register-austria.at
Email info@capri-register-austria.at

Capri Schweiz
www.ford-capri.ch
Email info@ford-capri.ch

Ford Capri Club Nederland
www.fordcapriclubnederland.nl
Email info@fordcapriclubnederland.nl

Worldwide clubs

Australian Club
Capri Car Club Inc
www.capricarclub.org.au
Email membership@capricarclub.org.au

New Zealand Club
Capri Car Club of NZ
www.fordcapri.co.nz
Email roncapri@flexolink.com

US/Canadian Club
Capri Club North America
www.capriclub.com
Email V6capri@aol.com

Suppliers are listed in alphabetical order. Inclusion is not a recommendation; please treat any omissions as accidental. The best solution is to speak to owners in your area for up to date information on the best business to satisfy your requirements.

European parts suppliers

Aldridge
(trim)
www.aldridge.co.uk

Capri Gear
www.caprigear.co.uk

Car Parts Direct
www.carparts-direct.co.uk

Classic Ford Parts
Email classicfordparts@hotmail.com

Classic Reproductions
(Engine bay stickers etc)
www.classicrepro.co.uk

East Kent Trim Supplies
(windscreen and other rubbers)

Ex-pressed Panels
www.steelpanels.co.uk

Hadrian Panels
www.carpanels.co.uk

Honeybourne Mouldings
(Glass fibre wings).
www.honeybournemouldings.co.uk

MWR Capri
www.mwrcapri.co.uk

Old Cars (Motormobil)
www.oldcars.de

Smith and Deakin
(Glass Fibre wings)
www.smithanddeakin.co.uk

Specialised Engines
www.specialisedengines.co.uk

Super Flex
(Polyurethane bushes)
www.superflex.co.uk

Tickover
www.tickover.co.uk

WST
www.capri-wst.de

US parts suppliers

Team Blitz
www.teamblitz.com

Australian parts

Capri Clinic
www.capriclinic.com

Books

Ford Capri
Graham Robson
ISBN 978-1861269782 (Crowood Autoclassics)

Capri Workshop Manual (Original)
ISBN 978-1855206878 (Brooklands Books)

Capri Mk2 1.6 and 2-litre Workshop Manual
Haynes and Ward
ISBN 978-1850103134 (Haynes Manuals)

Ford Capri Mk2 V6 Models
Legg
ISBN 978-1850103097 (Haynes Manuals)

Ford Capri Restoration Manual
Henson
ISBN 978-1844251124 (G T Foulis)

Magazines

The Capri doesn't have a large enough following to have a dedicated mainstream magazine, so it's a matter of looking at classic car publications in general each month to see if they're running a feature. The following periodicals are solely Ford orientated, though some include more tuning and modification articles, so not necessarily for the purist.

Classic Ford Magazine (UK)
www.classicfordmag.co.uk

Fast Ford Magazine (UK)
www.fastfordmag.co.uk

Legendary Ford Magazine (USA)
www.legendaryfordmagazine.com

Mustangs & Fords (USA)
www.mustangandfords.com

Performance Ford Magazine (Aus and NZ)
www.performanceford.com.au

Performance Ford Magazine (UK)
www.performancefordmag.com

Retro Ford Magazine (UK)
www.retrofordmag.co.uk

Street Fords (Aus)
www.streetfords.com/

17 Vital statistics
– essential data at your fingertips

Technical specifications of selected cars from each decade

The figures expressed below are only an indication of engine output and performance, averaged from several sources. Output varied throughout production: for example, GT versions were always more powerful than base specification cars, with the same capacity motor. Later versions of all models became more powerful with the passage of time.

The '60s Mk1 1600 UK, 4-cyl OHV in-line engine, 64bhp@5000rpm, torque 91lb/ft@2500rpm, 90mph top speed, 28mpg
Mk1 2-litre V4 UK, 4-cyl OHV 60 degree V4, 92bhp@5250rpm, torque 104lb/ft@4000 rpm, 105mph top speed, 25mpg

The '70s Mk1 3-litre V6 UK, 6-cyl OHV V6, 128bhp@4750rpm, torque 192lb/ft@4000rpm, 120mph top speed, 20mpg
Mk2 1600 UK, 4-cyl OHC in-line engine UK, 72bhp@5500rpm, torque 87lb/ft@2700rpm, 95mph top speed, 30mpg

The '80s Mk3 2-litre UK, 4-cyl OHC in line engine, 98bhp@5500rpm, torque 111lb/ft@3500rpm, 105mph top speed, 25mpg
Mk3 2.8i, 6-cyl V6 OHV, 160bhp@5700rpm, torque 152lb/ft@4000rpm 131mph top speed, 23mpg

Timeline: the major events in the development of the Capri

1969 The Capri is launched at the Brussels Motor Show. In the UK the car will have 1300 and 1600 X Flow engines or the Essex 2-litre V4; Germany will have Cologne V4s in 1300, 1500, 1700 and 2-litre capacities.
1969 The 2.3-litre Cologne V6 is launched in Germany.
1969 The Essex 3-litre V6 arrives.
1970 The 3-litre E version launched in the UK as the top of the range car with all the trimmings.
1970 The Capri hits America fitted with the 1600 X Flow motor. The cars were assembled in Germany.
1970 The Germans produce a 2.6-litre V6.
1972 The 'facelift' models arrive, suspension changes, power bulge bonnet, rectangular headlights, larger back light units and the Pinto engine for the 1600 mark the main differences.
1973 One million Capris made.
1974 The Mk2 version is produced. Now a hatchback, the car is now longer, wider and heavier than its predecessor.
1975 The Capri gets a full 12 month manufacturer's warranty. Power steering on the 3-litre Ghia.
1976 All Capris are now built in West Germany.
1978 The Mk3 is launched. Car receives aerodynamic body updates and an

improved interior. Now has 4 headlights to replace the two rectangular ones fitted previously.

1979 Automatic chokes arrive along with viscous fans.

1981 The 2.8i fitted with Bosch K Jetronic replaces the 3-litre cars. Wide wheels, vented discs, gas shocks and uprated anti-roll bars help contain the extra power.

1983 5-speed box for the 2.8i

1983 Range reduced to a slimline 1.6LS, 2.0S and the 2.8i.

1984 Laser introduced as a special edition but proves so popular that it becomes a mainstream product.

1984 Only the Laser in 1600 and 2-litre versions plus the new 2.8 Injection Special remain on sale.

1984 Left-hand drive production ends

1986 1038 special edition Capri 280s end the production of the Capri with more than 1.8 million built.

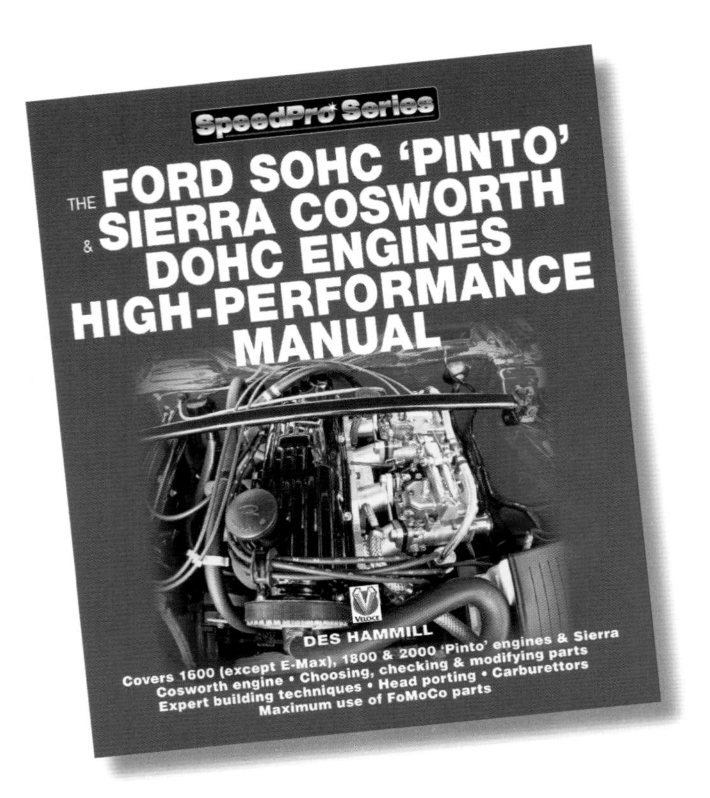

Expert practical advice from an experienced race engine builder on how to build a high-performance version of Ford's naturally aspirated 4-cylinder 1600, 1800 & 2000cc Pinto engine, which has been used in Ford's most popular cars (Escort, Capri, Cortina & Sierra – Ford/Mercury Capri, Pinto, Bobcat in USA) over many years. Whether the reader wants a fast road car or to go racing, Des explains, without using technical jargon, just how to build a reliable high-power engine using as many stock parts as possible and without wasting money on parts and modifications that don't work. Also covers Cosworth versions of Pinto engines and fitting Cosworth heads to Pinto blocks. Does not cover 1300, E-Max 1600 or American-built 2300.

• Paperback • 25x20.7cm • £19.99
• 144 pages • 200+ b&w pictures
• ISBN: 978-1-903706-78-7
• UPC: 6-36847-00278-7

Index